Before Galileo

Before Galileo

The Birth of Modern Science in Medieval Europe

JOHN FREELY

OVERLOOK DUCKWORTH
NEW YORK · LONDON

This edition first published in hardcover in the United States and the United Kingdom in 2012 by
Overlook Duckworth, Peter Mayer Publishers, Inc.

NEW YORK
141 Wooster Street
New York, NY 10012
www.overlookpress.com
For bulk and special sales, please contact sales@overlookny.com

LONDON
90-93 Cowcross Street
London EC1M 6BF
inquiries@duckworth-publishers.co.uk
www.ducknet.co.uk

Cataloging-in-Publication Data is available from the Library of Congress

A catalogue record for this book is available from the British Library

Book Design and typeformatting by Bernard Schleifer
Manufactured in the United States of America

ISBN US: 978-1-59020-607-2
ISBN UK: 978-0-7156-4510-9

1 3 5 7 9 10 8 6 4 2

For my beloved Toots

Contents

Before Galileo

Introduction

HEN I FIRST STARTED TEACHING PHYSICS, THE STANDARD narrative was that modern science began with the heroic efforts of Galileo to gain acceptance for the revolutionary sun-centered worldview of Copernicus, as opposed to the ancient geocentric cosmology of Aristotle and Ptolemy accepted by academia and the Church. It was this crusade that sparked the Scientific Revolution and culminated in the new physics and astronomy of Newton, at the dawn of the modern scientific age.

Virtually nothing was said in this narrative about the predecessors of Copernicus, Galileo, and Newton, although historians of medieval European civilization have in recent years traced back the beginning of science in the West more than a millennium before them. In reality, an impressive succession of European scholars opened the way for the Scientific Revolution, laying the foundations for the breakthrough theories and discoveries later made some of which they anticipated.

Before Galileo seeks to right this historical injustice, something I started thinking about while studying physics in college. I did my undergraduate studies on the GI Bill after World War II at Iona College in New Rochelle, New York, founded by the Irish Christian Brothers. The first thing I noticed on the campus was a statue of Saint Columba, patron saint of the Irish Christian Brothers who lived in the sixth century. Columba had been forced out of Ireland and founded a monastery on the west coast of Scotland on the island of Iona, the legendary burial place of Macbeth. His students went on to found other monastic schools in England and then on the Continent, beginning the reeducation of western Europe in the Dark Ages, just as the Irish Christian Brothers who taught me had founded a college in a suburb of New York City, which they may have felt was in need of enlightenment. I certainly felt the need, for I had dropped out of school at seventeen to join the U.S. Navy, as had a number of my classmates.

My physics teacher at Iona was Brother Thomas Bullen, who had studied physics with P. M. S. Blackett, winner of the Nobel Prize in Physics in 1948. I knew that Blackett had studied at Cambridge under Lord Rutherford, the founder of nuclear physics, who was awarded the Nobel Prize in Chemistry in 1908. Rutherford in turn had studied at Cambridge with James Clerk Maxwell, the father of modern electromagnetic theory. Subsequently, with the aid of the Math-Physics Genealogy website, I was able to trace my scientific ancestry through Brother Bullen, Blackett, Rutherford, and Maxwell in an unbroken line that included Newton, Leibniz, Galileo, Copernicus, and on back to the first Greeks who graduated from Italian universities, and through them to George Gemistus Plethon, who graduated from the University of Constantinople circa 1375 and was the principal source in bringing Greek learning to Italy, giving rise to the Italian Renaissance. That the link was unbroken all the way to antiquity intrigued me. Who carried the torch in those dark years before Europe's rebirth?

The most direct influence on this book, however, came from my postdoctoral study at All Souls College at Oxford with Alistair Cameron Crombie, renowned for his pioneering research in the history of medieval European science. After that, in addition to my courses in history and astronomy at what is now the University of the Bosphorus in Istanbul, I began teaching a course called The Emergence of Modern Science, East and West, a large part of which was based on what I had learned from Crombie, to which I have kept adding material on medieval European science up to the present day.

The principal idea that I inherited from Crombie was the continuity of western European science from the Dark Ages up through the times of Copernicus, Galileo, and Newton. More recent historians have questioned this notion. Thomas Kuhn's *The Copernican Revolution* (1957) and *The Structure of Scientific Revolutions* (1962) emphasized the paradigm shift involved in the heliocentric theory as evidence of a discontinuity of post-Copernican science with the scientific tradition that had developed in western Europe during the Middle Ages. Kuhn certainly has a point, but as Crombie wrote of Copernicus in his *Medieval and Early Modern Science*, first published in 1952: "He is a supreme example of a man who revolutionized science by looking at the old facts in a new way." Crombie goes on to point out the theories and data that Copernicus had inherited from his medieval European and ancient Greek predecessors, which is what I have done more thoroughly here, adding the contributions made by Islamic and Byzantine scientists.

Before Galileo begins with a look at western Europe at the beginning of the Dark Ages, with the Visigoth sack of Rome in 410 and the burning soon afterward of the great Library of Alexandria, with its vast collection of all the Greek works from those of Homer onward, as the ancient Graeco-Roman world was coming to an end in the gathering darkness of the early medieval period.

The Alexandrine Library contained copies of all the works of Greek science from the Pre-Socratics through the great mathematical physicists and astronomers of the Hellenistic period. Socrates himself wrote nothing, but he taught Plato, who in turn taught Aristotle, who taught Theophrastus, and so on, starting the chain of teacher and student, which was broken by the collapse of classical civilization and the burning of the Library in Alexandria, with the loss of all of their works.

But a number of the classics of Greek science and philosophy survived through a tenuous Ariadne's thread that wound its way from Alexandria through the medieval Byzantine and Islamic worlds, involving, in the latter case, translations from Greek to Aramaic to Persian to Arabic, and then eventually into Latin.

Before these Latin translations became available in western Europe, only a few remnants of classical learning were preserved by increasingly isolated Roman scholars, most notably Boethius and Cassiodorus. But more substantial remains of classical learning made their way to the first Irish monasteries, principally those of Saint Columba, where a number of Greek-speaking scholars took refuge, crossing with him to Iona, beginning the reeducation of Europe and bringing light to the Dark Ages. Eventually this reeducation reached a high enough level for European scholars to understand Graeco-Arabic science in Latin translation, a process that accelerated with the founding of the first European universities in the twelfth and thirteenth centuries.

But in the earlier medieval period European scholars had to start literally from scratch, driven by curiosity and observation of the world around them and the heavens above. Thus in the process Western science had from the very beginning a quality of practical empiricism that distinguishes it from the more abstract character of most Greek and Islamic science. This is evident in the work of Newton who, as Crombie wrote, "achieved the clearest appreciation of the relation between the empirical elements in a scientific system and the hypothetical elements derived from a philosophy of nature."

We will see this quality of practical empiricism of the Venerable Bede, writing in the early eighth century, who noted that "we know, who live on the shore of the sea divided by Britain," how the wind could advance or retard a tide. Because of my early childhood in Ireland, I can relate to what Bede was saying. From age four to seven I lived with my mother's parents out on the Dingle peninsula in county Kerry, the westernmost point of Europe, where life was governed by the tides. My grandfather Tomas, an illiterate Irish-speaking fisherman, was known as Tom of the Winds because his seeming endless knowledge of the world was said to have been brought to him by the four winds. I always went with Tomas when he set his nets on the strand near our cottage, from where we could hear the rumbling of the potato-sized rocks as they rolled in and out with the rise and fall of the sea, and he would wet his forefinger and raise it to gauge the direction of the wind before setting out. Tomas was my first teacher, and it may have been in my talks with him that I began thinking about things like time and tide that eventually led me to write this book.

Whether you are a fisherman or a cobbler or a physicist, you need a teacher. That is what *Before Galileo* is all about—the transmission of knowledge from one person to another, which in the case of western Europe began during the long night of the Dark Ages, a thousand years before Galileo was born.

St. Jerome in his study, painting by Domenico Ghirlandaio, 1480

1

Light in the Dark Ages

AINT JEROME, IN A LETTER TO THE LADY PRINCIPIA IN AD 412, wrote that "a dreadful rumour has reached us from the West. We heard that Rome was besieged, that the citizens were buying their safety with gold, and that when they had been thus despoiled they were again beleaguered, so as to lose not only their substance but their lives." He went on to say: "The speaker's voice failed and sobs interrupted his utterance. The city which had taken the whole world was itself taken; nay, it fell by famine before it fell by the sword, and there were but few to be found to be made prisoner."

Jerome was bewailing the sack of Rome by the barbarian Visigoths under Alaric on August 24, 410. Worse was yet to come, for in 455, after the assassination of Valentinian III, Emperor of the West, Rome was sacked again by the barbarian Vandals, who plundered the undefended city for three days, inflicting far more damage than had Alaric. Victor of Vita, a North African bishop, reports on the shiploads of captives who were brought to Cyrenaica and sold in the slave markets there, leaving Rome virtually empty. Rome itself did not fall, but it was left in ruins and virtually unpopulated for several weeks, the institutions of government and education no longer functioning.

The Graeco-Roman world was coming to an end, overwhelmed by the onslaught of the "barbarians," its ancient gods and learning eclipsed by the rise of Christianity. The light of classical learning was also about to be extinguished in Alexandria, which—after its founding by Alexander the Great in 331 BC—had succeeded Athens as the intellectual center of the Greek world.

The Library at Alexandria, founded at the beginning of the fourth century BC, preserved everything written in Greek from the first edition of Homer, including the philosophical and scientific works of Plato, Hippocrates, Aristotle, Theophrastus, Hippocrates, Democritus, Epicurus, Euclid, Aristarchus, Archimedes, Eratosthenes, Apollonius, Hero, Hipparchus, Strabo, Ptolemy, Galen, Dioscorides, and Diophantus, to name only the most famous.

An imperial decree published in 391 by the Emperor Theodosius I, a Christian, ordered that all pagan temples and other institutions in the empire be closed, including the Library and Museum in Alexandria. The last head of the Library was the mathematician Theon Alexandricus (c. 335–405). Theon's daughter Hypatia, a distinguished philosopher and mathematician, was torn to pieces in March 415 by a mob of monks led by a zealot named Peter the Reader. Around the same time the Library was destroyed, one version of the story being that it was burned to the ground by fanatical Christians. In any event the Library had completely vanished by the early fifth century, and not a single one of the scrolls deposited there has survived. Those ancient Greek works that exist today, just a small fraction of the Library's original collection, are later copies, some in the original Greek, others in Arabic and Latin translations, preserved in medieval monasteries. Thus the burning of the Alexandrine Library meant the loss of works of Greek literature, history, and science created through a period of more than a thousand years; as the dying embers of these scrolls faded from sight, the long night of the Dark Ages descended upon the world.

The earliest scientific works that had been preserved at the Library in Alexandria are fragments by the so-called Pre-Socratics, who flourished during the last half of the Archaic period (c. 750–480 BC) and the beginning of the Classical era (479–323 BC), all of them either from the Aegean coast of Asia Minor or from Magna Graecia, the Greek colonies in southern Italy and Sicily. The first of them were Thales (c. 625–547 BC), Anaximander (c. 610–545 BC), and Anaximenes (fl. 546 BC), all of Miletus. Aristotle referred to them as *physikoi*, or physicists, from the Greek *physis*, meaning "nature" in its widest sense, contrasting them with the earlier *theologoi*, or theologians, for they were the first who tried to explain phenomena on natural rather than supernatural grounds. The Miletian physicists believed that all material things in nature were just different forms of an *arche*, or fundamental substance, which endured through all apparent change. Thales said that the *arche* was water, Anaximander thought that it was an undefined substance called *apeiron*, whereas Anaximenes held that it was *pneuma*, meaning "air" or "spirit." Thales undoubtedly chose water because at normal temperatures it is liquid, but when heated it becomes a gas, water vapor, and when frozen it becomes a solid, ice. Thus the same substance takes on different forms, depending on physical conditions. I often think of this in terms of my own self, because I am the same person I was when I was young, or at least I think so, despite the physical changes that have taken place. But I must say that I would be hesitant to meet my seventeen-year-old self; I would recognize him, but I wonder what he would think of how time would transform him.

Heraclitus of Ephesus (fl. c. 500 BC) believed that the enduring reality in nature is not Being, as in the existence of a fundamental substance, but Becoming, that is to say, perpetual change, which he expressed in his famous aphorism, "*Panta rhei*" ("Everything is in flux"). He gives an example in one of his surviving fragments, where he said, "You never step into the same river twice," meaning that not only has the river flowed on in the interim but he himself has changed.

The contrasting approaches of the Milesians and Heraclitus are both evident in the laws of physics that I teach in my physics classes. The Milesian view appears in laws such as that of the conservation of mass, which says that the total mass involved in a process is the same before and after a chemical reaction, though the masses of the individual constituents have changed. The approach of Heraclitus is taken in theories that focus on the time rate of change in quantities, such as Newton's second law of motion, which states that the acceleration of a body, that is, the time rate of change of its velocity, is equal to the force acting on it divided by his mass.

Pythagoras (c. 580–c. 500 BC) is known for his famous theorem, which says that in a right triangle the square erected on the hypotenuse has an area equal to the sum of the areas erected on the two sides, a relation that represents the beginning of Greek mathematics. Pythagoras and his followers are credited with doing the first experiments in physics, in which they studied the sounds made by musical instruments and discovered the numerical relations involved in musical harmony. This led them to believe that the cosmos itself was designed according to harmonious principles, which could be expressed in numerical relations similar to those they discovered in musical theory. The Pythagoreans went on to formulate a cosmology in which the five visible planets—Mercury, Venus, Mars, Jupiter, and Saturn—together with the sun, moon, and earth, rotated about the Hearth of the Universe, orbiting with velocities inversely proportional to their distance from the center, those that were closest and moving more rapidly giving more high-pitched sounds, and vice versa. This is the famous Music of the Spheres, which continued to fascinate up to the times of Copernicus, Kepler, and Shakespeare. Shakespeare, in *The Merchant of Venice*, has Lorenzo call Jessica's attention to the harmony of the celestial spheres:

> Sit, Jessica. Look how the floor of heaven
> Is thick inlaid with patines of bright gold;
> There's not the smallest orb which thou behold'st
> But in his motion like an angel sings,
> Still quiring to the young-ey'd cherubins.
> Such harmony is in immortal souls,
> But while this muddy vesture of decay
> Doth grossly close it in, we cannot hear it.

Whereas Heraclitus thought that everything was in a state of flux and nothing was permanent, Parmenides (c. 515–450 BC) believed that all Being is what he called the One, and denied absolutely the possibility of change.

He believed that the cosmos is a full (i.e., no void), uncreated, eternal, inde-structible, unchangeable, immobile sphere of being, and all sensory evidence to the contrary is illusory. One Parmenidean fragment stated, "Either a thing is or it is not," meaning that creation or destruction is impossible.

Echoes of this immutable Parmenidean cosmology reverberated from antiquity down to the European Renaissance, as in the last canto of Spenser's *The Faerie Queene*:

> Then gin I thinke on that which Nature sayd,
> Of that same time when no more Change shall be,
> But stedfast rest of all things, firmly stayd
> Upon the pillours of Eternity,
> That is contrayr to Mutabilitie;
> For all that moveth doth in Change delight.

A way out of the Empedoclean impasse was provided by the atomic the-ory of Leucippus (fl. early fourth century BC) and his more famous pupil, Dem-ocritus (c. 470–c. 404 BC). They held that the *arche* existed in the form of atoms, the irreducible minima of all physical substances, which in their ceaseless motion through the void collide and combine with one another in various con-figurations to take on all of the innumerable forms observed in nature. One of the extant fragments of Leucippus says, "Nothing occurs at random, but every-thing for a reason and by necessity," by which he meant that atomic motion is not chaotic, but obeys the immutable laws of nature. The atomic theory was not generally accepted in the time of Democritus, largely because of its deter-ministic character, for it allows no chance, choice, or free will.

Some of the profound questions raised by Parmenides were addressed by Empedocles (c. 482–443 BC). While Empedocles agreed with Parmenides that there was a serious problem regarding the reliability of sense impressions, he felt that since our senses were the only direct contact with the world of na-ture, we could still make use of them through cautious evaluation of the in-formation they provided. He tried to address the problem of change by saying that there is not one fundamental *arche* but four—earth, water, air, and fire

—which generate all the material substances in nature by mixing together in various ways under the influences of forces he called Love and Strife.

Anaxagoras of Clazomenae (c. 500–428 BC) postulated another element called the aether, which was in constant rotation and carried with it the celestial bodies. He also believed that there was a directing intelligence in nature that he called *Nous*, which gives order to what otherwise would be a chaotic universe. By *Nous* he meant literally "the Mind of the Cosmos," just as in a very real sense our own minds give order to the world around us.

Anaxagoras was the last of the Ionian physicists, for when he came of age he moved to Athens, which at the beginning of the Classical period in 479 BC emerged as the political and intellectual center of the Greek world. He was the first philosopher to live in Athens and became the teacher and close friend of the great Athenian statesman Pericles (c. 495–429 BC).

When Pericles delivered his famous funeral oration in 431 BC, honoring the Athenians who fell in the first year of the Peloponnesian War, he reminded his fellow citizens that they were fighting to defend a free and democratic society that was "open to the world," one whose "love of the things of the mind" had made their city "the school of Hellas." "Mighty indeed," he said, "are the marks and monuments of our empire which we have left. Future ages will wonder at us, as the present age wonders now."

The most famous of the Athenian schools was the Academy, founded soon after 386 BC by Plato (c. 428–c. 347 BC), who had been a student of Socrates (c. 470–399 BC). I should say that Plato was one of the young men who conversed with Socrates when the sage held forth in the Agora, the marketplace of Athens. The Academy was one Attic mile from the Dipylon Gate in the walls of ancient Athens along the Sacred Way that led to Eleusis. A large part of the site has been excavated, and when we were living in Athens I walked there one day, following the route of the bus route marked AKADEME. This reminded me of the lines in *Paradise Regained* where Milton describes the school as "the olive grove of Academe, Plato's retirement, where the Attic bird trills her thick-warbl'd notes the summer long."

Plato believed that mathematics was a prerequisite for the dialectical process that would give the future leaders educated at the Academy the philosophical insight necessary for governing the ideal state, which he describes in the *Republic*. Plato's most enduring influence on science was his advice to approach the study of nature as an exercise in geometry. Through this "geometrization of nature," which could best be done in disciplines that could be suitably idealized, such as astronomy, one can formulate laws that are as "certain" as those in geometry. As Plato has Socrates remark in the *Republic*: "Let's study astronomy by means of problems, as we do geometry, and leave the things in the sky alone."

The principal problem in Greek astronomy was to explain the motion of the celestial bodies—the stars, sun, moon, and the five visible planets—as seen from the earth. The celestial bodies all seem to rotate daily about a point in the heavens called the celestial pole. According to the heliocentric theory of Copernicus, the celestial pole is actually the projection of the earth's north pole among the stars, and its apparent motion is actually due to the axial rotation of the earth in the opposite sense. Although the sun

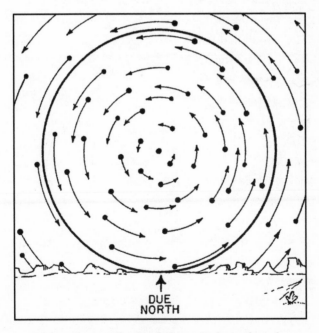

DUE
NORTH

The apparent motion of typical stars in the northern sky

rises in the east and sets in the west each day, from one day to another it appears to move back from west to east about 1°, making the transit of the twelve signs of the zodiac in one year, an apparent motion produced by the orbiting of the earth around the sun, which the Greeks, whose cosmology was geocentric, explained through complicated mathematical theories.

The apparent path of the sun through the zodiac, the so-called ecliptic, makes an angle of about 23.25° with the celestial equator, the projection of the earth's equator among the stars. This is due to the fact that the earth's axis is tilted by that angle with respect to the perpendicular of the ecliptic plane, an obliquity that is responsible for the recurring cycle of seasons.

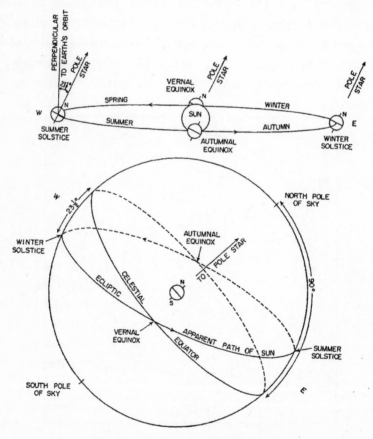

The tilt of the Earth's axes as the cause of seasons

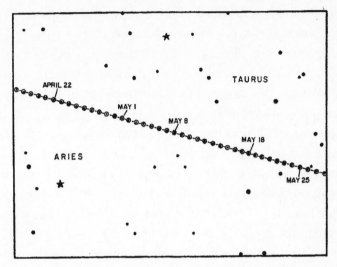

Above: The apparent motion of the sun through the constellations Aries and Taurus

Below: The apparent motion of Mars through Aries and Taurus,
showing its retrograde motion

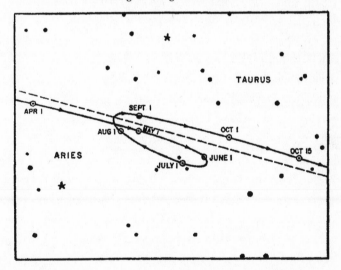

The planets all move in paths that are close to the ecliptic, moving from east to west during the night along with the fixed stars, while from one night to the next they generally move slowly back from west to east around the zodiac. Each of the planets also exhibits a periodic retrograde motion, which

· 25

shows as a loop when its path is plotted on the celestial sphere. This is due to the fact that the earth is moving in orbit around the sun, passing the slower outer planets and being itself passed by the swifter inner planets, the effect in both cases making it appear that the planet is moving backward for a time among the stars. This mysterious motion compelled the Greeks toward scientific discovery, to give order to celestial motion.

Spurring on his students, Plato posed a problem: to demonstrate "on what hypotheses the phenomena (i.e., the apparent retrograde motions) concerning the planets could be accounted for by uniform and ordered circular motions." Eudoxus of Cnidus (c. 400–c. 347 BC), a younger contemporary of Plato at the Academy, was the first to try his hand at a solution. Eudoxus was the greatest mathematician of the classical period, credited with some of the theorems that would later appear in the works of Euclid and Archimedes. He was also the leading astronomer of his era and had made careful observations of the celestial bodies from his observatory at Cnidus, on the southwestern coast of Asia Minor. Eudoxus suggested a complicated mechanical model known as the theory of homocentric spheres, which successfully reproduced the apparent retrograde motion of the planets, though it had no basis in any physical theory.

Meanwhile the renowned physician Hippocrates (c. 460–c. 370 BC) had established a school of medicine on the island of Kos. The writings of Hippocrates and his followers, the so-called Hippocratic corpus, comprising some seventy works dating from his time to about 300 BC, represents the beginning of Greek medicine as a science. Beside treatises on the various branches of medicine, they include clinical records and notes of lectures given to the general public on medical topics. A treatise on deontology, or medical ethics, contains the famous Hippocratic oath, which is still taken by physicians today. One work in the Hippocratic corpus is entitled *The Sacred Disease*, for the name given to epilepsy, since those suffering from it were believed to be stricken by the gods. The author of this work says that epilepsy, like all other diseases, has a natural cause, and those who first called it sacred were trying to cover up their ignorance. "I do not believe," he said, "that the 'Sacred Disease' is any more divine or sacred than any other disease but, on the contrary, has specific characteristics and a definite cause. Nevertheless, because it is completely different from other diseases, it has been regarded as a divine visitation by those who, being only human, regard it with ignorance and astonishment."

Plato's most renowned student was Aristotle (384–322 BC), who in 335 BC founded in Athens a school called the Lyceum, on the site of what is now the Greek Parliament, which rivaled the Academy in its fame. Aristotle's writings are encyclopedic in scope, including works on logic, metaphysics, rhetoric, theology, politics, economics, literature, ethics, psychology, physics, mechanics, astronomy, meteorology, cosmology, biology, botany, natural history, and zoology.

The dominant concept in Aristotle's philosophy of nature is his notion of causation. When looking for the cause of anything we must distinguish between the following four causes, identified by Aristotle as material, formal, efficient, and final. The material cause of something is known when we identify the material out of which it is formed; the formal cause when we specify the plan or design according to which it is fashioned; the efficient cause when we name the agent that actually made it; the final cause when we give the reason why it was brought into being.

The first three causes—material, formal, and efficient—correspond to the three aspects of existence that can be distinguished as matter, form, and actualization of form. But these do not define the course of events in nature, since an acorn, for example, always grows into an oak and never a cypress. The final cause states that each substance has an inherent purpose. Thus there must be a purpose or design in the acorn such that it always grows into an oak tree. This aspect of existence is indicated by the word *entelechy*; this means the purpose that guides things to develop in one way rather than another.

The main outlines of Aristotle's cosmology were inherited from earlier Greek thought, which distinguished between the imperfect and transitory terrestrial region below the sphere of the moon and the perfect and eternal celestial region above. He took from the Milesian physicists the notion that there was one fundamental substance in nature and reconciled this with

Empedocles' concept of the four terrestrial elements—earth, water, air, and fire—to which he added the aether of Anaxagoras as the basic substance of the celestial region.

Aristotle's cosmology arranged the four elements in order of density, immobile spherical earth at the center, surrounded by concentric shells of water (the ocean), air (the atmosphere), and fire, the latter including not only flames butextraterrestrial phenomena such as lightning, rainbows, and comets. The natural motion of the terrestrial elements was to their natural place, so that if earth is displaced upward in air and released, it will fall straight down, whereas air in water will rise, as does fire in air. This linear motion of the terrestrial elements is temporary since it ceases when they reach their natural place.

Aristotle also tried to explain why a projectile continues to move when it is no longer in contact with its motive force. His ingenious but incorrect explanation involves a hypothetical phenomenon called antiperistasis, in which the air displaced by the front of the projectile moves into the temporary partial vacuum in its wake and gives it a forward impetus. These three erroneous theories—that the velocity of a falling body is proportional to its weight, that a void is impossible, and the notion of antiperistasis—persisted for more than a millennium, until they were refuted by the new dynamics that was developed by medieval European scholars, culminating in the laws of motion formulated in 1687 by Isaac Newton.

According to Aristotle, the celestial region begins at the moon, beyond which are the sun, the five planets, and the fixed stars, all embedded in crystalline spheres rotating around the immobile earth. The celestial bodies are made of aether, the quintessential element, whose natural motion is circular at constant velocity, so that the motions of the celestial bodies, unlike those of the terrestrial region, are unchanging and eternal. Aristotle used Eudoxus's theory of the homocentric spheres to create a physical model of his world picture. He added a number of "counteracting spheres" to the homocentric theory of Eudoxus, so as to unify the motion of all the celestial bodies.

Aristotle's Cosmology, from Petrus Apianus,
Cosmographia per Gemma Phrysius Restitua, *Antwerp, 1539*

Heraclides Ponticus (c. 390–c. 339 BC), so called because he was a native of Heraclea on the Pontus (the Black Sea), was a contemporary of Aristotle and had also studied at the Academy under Plato. His cosmology differed from that of Plato and Aristotle in at least two fundamental points, which may be due to the fact that after leaving the Academy he seems to have studied with the Pythagoreans. The first point of difference concerned the extent of the cosmos, which Heraclides thought to be infinite rather than finite. A second difference involved the apparent circling of the stars around the celestial pole, which Heraclides said was actually due to the rotation of the earth on its axis in the opposite sense. This revolutionary theory, revived nineteen centuries later by Copernicus, was generally rejected by the Greeks, since it ran counter to the static geocentric model of Aristotle and his successors.

Aristotle was succeeded as head of the Lyceum by his student Theophrastus (c. 371–c. 287 BC), to whom he bequeathed his enormous library, which included copies of all his works. Theophrastus is considered to be the second founder of the Lyceum, which he directed for thirty-seven years, reorganizing and enlarging the school.

Theophrastus was as prolific and encyclopedic as Aristotle, having written 227 books, most of which are now lost. Two of his extant works, the *History of Plants* and the *Causes of Plants*, have earned him the title Father of Botany, while his book *On Stones* represents the beginning of geology and mineralogy. His work on human behavior, entitled *Characters*, is a fascinating description of the types of people living in Athens during his time, all of whom still seem to be represented in modern Athens, as I observed when we were living there.

Two other schools of philosophy founded in Athens late in the fourth century BC were not formal institutions like the Academy and the Lyceum, but more loosely organized groups gathered to discuss philosophy. One of the schools, known as the Garden, was founded by Epicurus of Samos (341–270 BC) and the other, the Porch, was begun by Zeno of Citium (c. 335–c. 263 BC), each school named for the meeting place favored by its founder. Both Epicurus and Zeno created philosophical systems that were divided into three parts—ethics, physics, and logic—in which the last two were subordinate to the first, whose goal was to secure happiness.

The physics of Epicurus was based on the atomic theory, to which he added one new concept, that an atom moving through the void could at any instant "swerve" from its path. This eliminated the absolute determinism that had made the original atomic theory of Leucippus and Democritus unacceptable to those who, like the Epicureans, believed in free will. Zeno and his followers, who came to be known as the Stoics, rejected the atom and the void, for they looked at nature as a continuum in all of its aspects—space, time, and matter—as well as in the propagation and sequence of physical phenomena. These two opposing schools of thought about the nature of the cosmos—the Epicurean atoms in a void versus the continuum of the Stoics—have competed with one another from antiquity to the present, for they seem to represent antithetical ways of looking at physical reality.

Early in the Hellenistic period, which began with the death of Alexander the Great in 323 BC, the center of the Greek intellectual world shifted from Athens to Alexandria, the new city that had been founded by Alexander on the Canopic branch of the Nile. Alexandria was the capital of the Ptolemaic kingdom, founded by Ptolemy I (r. 305–281 BC), one of Alexander's generals. Ptolemy sought to make his city a great cultural center by founding a research institution known as the Museum, since it was dedicated to the Muses, with a Library that he and his immediate successors in the Ptolemaic dynasty stocked with all of the works of Greek authors from Homer onward.

The Alexandrian Museum and its associated Library were designed as a school of higher studies, patterned on the famous schools of philosophy in Athens, most notably the Academy and the Lyceum. The Museum was a center of research in the tradition of the Lyceum of Aristotle and Theophrastus, with an emphasis on science. The scientific character of the Museum was probably due to Strato of Lampsacus (c. 340–c. 270 BC), who moved to Alexandria around 300 BC, to serve as tutor to the future Ptolemy II Philadelphus, remaining there until he returned to Athens to succeed Theophrastus as director of the Lyceum. The prince developed a deep interest in geography and zoology through his studies with Strato, and this was manifested in the development of the Museum when he succeeded his father as king in 283 BC.

The organization of the Library was probably due to Dimitrios of Phaleron (c. 350–c. 285 BC), the former governor of Athens, who had been forced to flee from Athens in 307 BC, after which he was given refuge in Alexandria by Ptolemy I. Dimitrios, who had been a student of Theophrastus at the Lyceum, is believed to have been the first chief librarian at the Library, a post he held until 284 BC. According to Aristeas Judaeus, a Jewish scholar in the reign of Ptolemy II, Dimitrios "had at his disposal a large budget in order to collect, if possible, all the books in the world, and by purchases and transcriptions he, to the best of his ability, carried the king's objective into execution."

This policy continued through the reigns of Ptolemy II and Ptolemy III Euergetes (r. 247–221 BC). Athenaeus of Naucratis reports that Ptolemy II bought the books of Aristotle and Theophrastus and transferred them to "the beautiful city of Alexandria." By the time of Ptolemy III the Library was reputed to have a collection of more than half a million parchment rolls.

The third Ptolemy built a new branch of the Library within the Serapeum, the temple of Serapis. Ephanius of Salamis, a Christian scholar of the fourth century AD, refers to this addition in writing of the "first library and another built in the Serapeum, smaller than the first, which was called the daughter of the first."

Dimitrios was succeeded as chief librarian by Zenodotus of Ephesus, who held the post until 245 BC. His principal assistant was the poet Callimachus of Cyrene (c. 305–c. 240 BC), who classified the 120,000 works in the library according to author and subject, the first time that this had ever been done. His compilation, known as the *Pinakes* (*Tables*), was entitled *Tables of Persons Eminent in Every Branch of Learning Together with a List of Their Writings*, and filled more than 120 books, five times the length of Homer's *Iliad*.

Most of the great scientists of the Hellenistic era were associated with the Museum and the Library. Eratosthenes (c. 276–c. 194 BC), who was appointed chief librarian of the Library around 235 BC, is renowned for his accurate measurement of the earth's circumference, which he computed to be 252,000 stades, about 20 percent different than the modern value.

The method that Eratosthenes used to measure the circumference of the earth involved simultaneous observations made at Alexandria and Syene, directly to the south on the same meridian. It was observed that at the summer solstice the sun was directly overhead at noon in Syene, while on a sundial at Alexandria it cast a shadow equal to one-fiftieth of a circle. Assuming that the sun is so far away that its rays at Syene and Alexandria are parallel, Eratosthenes concluded that the north-south distance between the two places is one-fiftieth of the earth's circumference. Thus the circumference of the earth is fifty times the distance between Syene and Alexandria, or 250,000 stades. This is equivalent to about 29,000 English miles, which may be compared with the modern figure for the earth's circumference of a little less than 25,000 miles.

Eratosthenes was also the first to draw a map of the known world using a system of meridians of longitude and parallels of latitude.

The great school of mathematics at Alexandria was founded by Euclid (fl. 295 BC), whose *Elements* laid the foundations not only of plane geometry but also of algebra and number theory. Euclid, still a towering figure in modern mathematics, established a logical form and order in which theorems are presented, which was to be the model for all future work in Greek mathematics and mathematical physics. Equally important is the axiomatic nature of the *Elements*, all of geometry following as a logical deduction from a few assumptions, themselves taken as necessarily true, which, when applied to physics and astronomy, represented the Platonic geometrization of nature.

A brilliant example of the geometrization of nature has survived in the treatise *On the Sizes and Distances of the Sun and Moon* by Aristarchus of Samos (c. 310–230 BC). Here the stellar and lunar sizes and distances were calculated from geometrical demonstrations based on three astronomical observations, together with an estimation of the earth's diameter. Aristarchus observed that the sun and the moon appear to be the same size, indicating that their diameters must be proportional to their distances from the earth. Second, he measured the lunar dichotomy—the angular separation of the sun and moon at half-moon—and then estimated the breadth of the earth's shadow where the moon passes through it at the time of a lunar eclipse. The results of these measurements led Aristarchus to conclude that the sun is about nineteen times farther from the earth than the moon, and that the sun is approximately 6.75 times as large and the moon about a third as large as the earth. All of his values are grossly underestimated because of the crudeness of his observations, but his geometrical methods were sound.

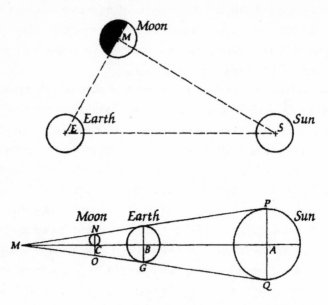

Geometrical demonstrations used by Aristarchus in his treatise
On the Sizes and Distances of the Sun and Moon

According to Archimedes, Aristarchus also wrote a book in which he proposed a theory in which the sun was the center of the cosmos, with the earth and the other planets orbiting around it. His contemporary Cleanthes of Assos (c. 331–c. 232 BC) wrote a tract condemning the heliocentric theory, stating that Aristarchus not only placed the earth in orbit around the sun but had it rotating on its axis. The heliocentric theory was not accepted in antiquity because it ran counter to the geocentric world-model formulated by Aristotle.

Greek mathematical physics reached its peak with the works of Archimedes (c. 287–212 BC), who was born in Syracuse in Sicily. Archimedes is said to have spent some time in Egypt, and he corresponded with Eratosthenes. It is probable that he studied in Alexandria under the successors of Euclid, for he was certainly familiar with the *Elements* and quoted from it extensively.

Apollonius of Perge (c. 262–c. 190 BC), a contemporary of Archimedes, wrote only one work that has survived, his treatise *On Conics*, and even there the last book is lost. This was the first comprehensive and systematic analysis of the three types of conic sections: the ellipse (of which the circle is a special case), parabola, and hyperbola.

In addition to this, Apollonius is credited with formulating mathematical theories to explain the apparent retrograde motion of the planets. One of the theories has the planet moving around the circumference of a circle, known as the epicycle, whose center itself moves around the circumference of another circle, called the deferent, centered at the earth. The second theory has the planet moving around the circumference of an eccentric circle, whose center does not coincide with the earth. He also showed mathematically that the epicycle and eccentric circle theories are equivalent, so that either model can be used to describe retrograde planetary motion.

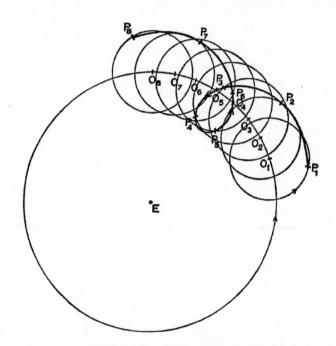

The epicycle theory used by Apollonius to explain the retrograde motion of the planets

Aside from the great theoreticians of the Hellenistic era, there were also a number of gifted inventors, whose works were extremely influential not only in technology and applied science but also in the theoretical development of pneumatics and even atomic theory. The most notable were Ctesibius of Alexandria (fl. 270 BC), Philo of Byzantium (fl. 250 BC), and Hero of Alexandria (fl. AD 62).

By far the greatest astronomer of antiquity was Hipparchus of Nicaea (c. 190–c. 120 BC), who flourished in the third quarter of the second century BC. Hipparchus is famous for his discovery of the precession of the equinoxes, that is, the slow movement of the celestial pole in a circle about the perpendicular to the ecliptic. The earth's precession manifests itself as a gradual advance of the spring equinox along the ecliptic, thus causing a progressive change in the celestial longitude of the stars. Hipparchus discovered this effect by comparing his star catalogue with observations made 128 years earlier by the astronomer Timocharis (c. 320–260 BC). Meticulously examining the differences, he concluded that the celestial longitude of the star Spica in the constellation Virgo had changed by two degrees in that interval of time. This amounts to an annual precession of 45.2 seconds of arc, which means that the axis of the earth describes a great circle in the celestial sphere in a period of twenty-six thousand years, a result verified by Newton in 1687. Hipparchus is also celebrated as a mathematician, his great achievement being the development of spherical trigonometry, which he applied to problems in astronomy.

Greek astronomy culminated with the work of Claudius Ptolemaeus (c. 100–c. 170), known more simply as Ptolemy. The most influential of his writings is his *Mathematical Synthesis*, better known by its Arabic name, the *Almagest*, the most comprehensive work on astronomy that has survived from antiquity. Here Ptolemy took the observational data of Hipparchus, along with his own observations, and converted it into the numerical parameters for his planetary models. He then used these models to construct tables [*ephemerides*] from which the solar, lunar, and planetary positions can be calculated for any given time in the future, as well as eclipses of the sun and moon. His planetary models used the epicycles and eccentric circles of

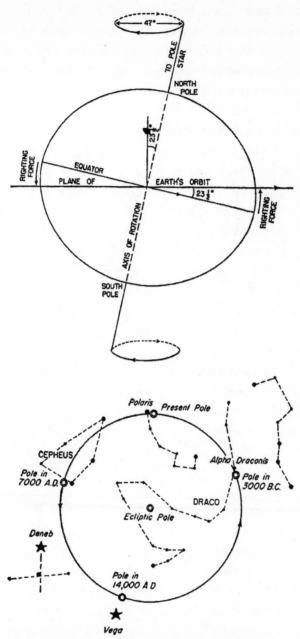

*The slow continuous change in the axis of Earth's rotation
is referred to as the precession of the equinoxes*

Apollonius, along with the spherical trigonometry of Hipparchus. The principal modification made by Ptolemy is that the center of each epicycle moves uniformly with respect to a point called the equant, which is displaced from the center of the deferent circle, a device that came to be the subject of controversy in later times.

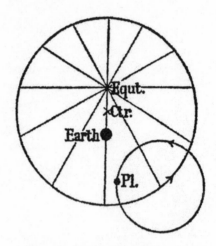

Above: Ptolemy's concept of the equant
Below: A simplified model of Ptolemy's planetary model

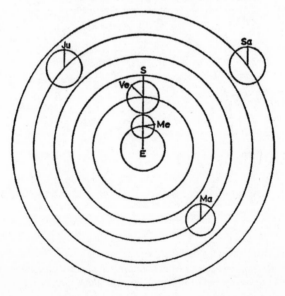

Ptolemy was a renaissance man even before the term had meaning. During his lifetime he wrote treatises on astrology, optics, geography, and musical theory, as well as making important advances in mathematics in his astronomical and other works that would be extremely influential in the development of science in both the Islamic world and western Europe.

In *Optics*, his treatise on light, Ptolemy gives the correct form for the law of reflection, already known to Euclid, which states that the incoming and reflected rays make the same angle with the surface of the mirror. His experiments led him to formulate an empirical relation for the law of refraction, the bending of light when it passes from one medium to another. He found that when light passes into a denser medium, as from air to water or glass, the refracted ray makes a smaller angle with the surface normal than does the incident ray. He then used these laws to find the location, size, and form of images produced by reflection and refraction.

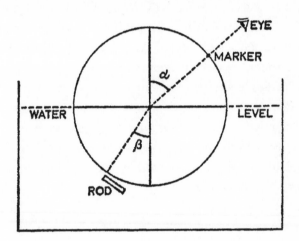

Ptolemy's experimental investigation of refraction

Ptolemy's *Geography* is the most comprehensive work on that subject to survive from the ancient world. One major defect of his work is that his value for the circumference of the earth is too small by a factor of roughly one-third. Additionally, in his map of the known world is the extension of the Eurasian landmass over 180° of longitude instead of 120°. Nevertheless, Ptolemy's treatise was by far the best geographical work produced in antiquity.

Galen of Pergamum (130–after 204), the most renowned medical writer of antiquity, was a younger contemporary of Ptolemy. Galen's writings, translated successively into Arabic and Latin, served as the basis for explaining the anatomy and physiology of the human body up until the seventeenth century, earning him the title Prince of Physicians. The title of one of his treatises is *That the best doctor is also a philosopher*. His philosophical bent is evident in his medical writings, where he interprets the work of Plato, Aristotle, Epicurus, and others, and also in his treatises *On Scientific Proof* and *Introduction to Logic*. He wrote on psychology as well, including an imaginative analysis of dreams, seventeen centuries before Freud. One of the psychic complaints recognized by Galen was lovesickness, which he believed to be the principal cause of insomnia, and he noted that "the quickening of the pulse at the name of the beloved gives the clue."

Much of Galen's pharmacological information comes from earlier sources, most notably Dioscorides (fl. AD 50–70), from Anazarbus in southeastern Asia Minor, who served a physician in the Roman army during the reigns of Claudius (r. 41–54) and Nero (r. 54–68). Dioscorides is regarded as the founder of pharmacology, renowned for his *De Materia Medica*, a systematic description of some six hundred medicinal plants and nearly a thousand drugs, including many we would rediscover later. One of the medicinal plants described by Dioscorides is cannabis, which he writes of in book 3 of *De material medica*:

> Cannabis: a plant useful for the plaiting of the most resilient ropes. It bears leaves like the ash but bad-smelling, has long hollow stems, and a round edible fruit which if eaten in quantity quenches sexual desire; pureed when green it's good for analgesic ear-drops. Wild cannabis bears shoots like those of the elm but darker and smaller; it's a cubit tall; the leaves are like the tame sort, rougher and darker, the flowers reddish like rose campion, and the seeds and roots like mallow.

The last great mathematician of antiquity was Diophantus of Alexandria (fl. c. AD 250), who did for algebra and number theory what Euclid had done for geometry. His most important treatise is the *Arithmetica*, of which six of the original thirteen books have survived. The six books of the *Arithmetica* that survived in Greek were translated into Latin in 1621, and six years later this work inspired the French mathematician Pierre Fermat to create the modern theory of numbers. After reading Diophantus's solution of the problem of dividing a square into the sum of two squares, which is the Pythagorean theorem, Fermat made a notation in the margin of his copy of the *Arithmetica*, stating, "It is impossible to divide a cube into two other cubes, or a fourth power or, in general, any number which is a power greater than the second into two powers of the same denomination." He then noted that he had discovered a remarkable proof, "which this margin is too narrow to contain." He never supplied the proof of what came to be called Fermat's last theorem, which was finally solved in 1994 by Andrew Wiles, a British mathematician working at Princeton University, the last link in a long chain of mathematical development that began in ancient Alexandria. The work of Diophantus is still a part of modern mathematics, studied under the heading of Diophantine analysis. An example is the problem of finding two numbers such that their sum is 20 and the sum of their squares 208, the answer being 8 and 12.

The last great philosopher of nature in the Greek world was John Philoponus (c. 490-570) of Alexandria, who criticized many of Aristotle's theories. He refuted Aristotle's theory that the velocities of falling bodies in a given medium are proportional to their weight, making the observation that "if one lets fall simultaneously from the same height two bodies differing greatly in weight, one will find that the ratio of the times of their motion does not correspond to the ratios of their weights, but that the difference in time is a very small one." Galileo drew the same conclusion more than a thousand years later in his famous demonstration at the Leaning Tower of Pisa.

He also criticized Aristotle's *antiperistasis* theory of projectile motion, which states that the air displaced by the object flows back to push it from behind. Instead Philoponus concluded that "some incorporeal kinetic power is imparted by the thrower to the object thrown" and that "if an arrow or a stone is projected by force in a void, the same will happen much more easily, nothing being necessary except the thrower." This is the famous "impetus theory," which was revived in medieval Islam and again in fourteenth-century Europe, giving rise to the beginning of modern dynamics.

After the destruction of the Library in Alexander all of the original works of Greek science in philosophy and science were lost, but copies of some of them survived and eventually made their way to western Europe by a number of routes and through various chains of translation.

Some fragments of Greek science survived in western Europe during the late Roman era. One of those responsible for this was Posidonius (fl. c. 135–51 BC), the last original thinker of the Stoic school, who studied in Athens before moving to Rhodes. There he attracted students from Rome, including Cicero, and was visited by leading Romans during their eastern journeys, most notable Pompey. He represented the Rhodians on an embassy to Rome in 87 BC, and his extensive travels took him as far as Gadeira (Cadiz) in Spain, where he observed the tides of the Atlantic.

Posidonius described the tides in his treatise *On Ocean*, where he ascribes the phenomenon to the combined actions of sun and moon. This treatise, based on his own observations and the works of Eratosthenes and Hipparchus, dealt with the entire globe. Eratosthenes had measured the meridional circumference of the earth to be 250,000 stades, but Posidonius disagreed with him and eventually settled on a value of 180,000 stades, which is about a third too low. Ptolemy used the lower value in his *Geography*, written c. AD 150, and this was the number used in the late fifteenth century by Christopher Columbus as the basis for his calculation of the distance between Iberia and India, reckoning, "If you sail from the west along the east wind you will reach India at a distance of 70,000 stades."

Posidonius also wrote extensively on history, geography, and both moral and natural philosophy, including commentaries on Plato's *Timaeus* and Aristotle's *Meteorology*. None of his wide-ranging works survive other than in fragments, though he was extremely influential in the subsequent intellectual history of the Roman world and of medieval Europe.

Another influential figure was Marcus Terentius Varro (116–27 BC), who studied in both Rome and Athens and wrote some seventy-five works on a wide range of subjects, almost all of which are lost. His most influential work was *Nine Books of Discipline*, which became the basis for all subsequent Latin encyclopedias. The nine disciplines referred to in the title were those that he identified as being necessary in the education of a Roman, namely grammar, rhetoric, logic, arithmetic, geometry, astronomy, musical theory, medicine, and architecture. Later writers left out the last two and the remainder came to be known as the seven liberal arts, which constituted the basis course of medieval schools in Europe, where the curriculum was known as the trivium (grammar, rhetoric, logic) and quadrivium (arithmetic, geometry, astronomy, musical theory).

Varro was a contemporary of the Roman poet Lucretius (c. 94–c. 50 BC), who was also influenced by Posidonius. Lucretius is renowned for a superb didactic poem in six books entitled *De rerum natura* (*On the Nature of Things*), based on the atomic theory and the teaching of Epicurus, as well as the commentary by Posidonius on Aristotle's *Meteorology*. The second book of *De rerum natura* deals with the kinetics of the atomic theory, with the random movements and collisions of the atoms bringing about the groupings and separations that form the various bodies found in nature. Lucretius followed Epicurus in saying that the reason the atoms collide is that "at quite indeterminate times and places they swerve ever so little from their course, just so much that you can call this a change in direction." This was later interpreted to mean that the atomic theory was not deterministic, since the "swerve" made atomic motion unpredictable. The atomic theory was in this way made more acceptable for Christians, since it allowed for free will in human actions. *De rerum natura* became popular in medieval Europe, which eventually led to the revival of the atomic theory in the seventeenth century.

The most important Latin encyclopedist and popularizer of science was Pliny the Elder (c. 23–79), who was born at Comum (Como) and was probably educated in Rome. He was in command of the Roman fleet sent to evacuate refugees during the eruption of Mount Vesuvius in AD 79, when he was asphyxiated by the volcanic fumes. His only surviving work is the *Natural History*, which historian Edward Gibbon described as an immense register in which the author has "deposited the discoveries, the arts, and the errors of mankind." Pliny noted in the preface "that by perusing about 2,000 volumes we have collected in 36 volumes about 20,000 noteworthy facts from one hundred authors we have explored." Pliny the Younger, his nephew and adopted son, described the *Natural History* as "a diffuse and learned work, no less rich in variety than nature itself." The *Natural History* was widely known in medieval Europe, where, despite its uneven quality and generally low level, it represented a large fraction of the scientific knowledge available at that time.

The Roman writer Seneca the Younger, a Stoic who flourished in the first century AD, is best known for his dialogues, letters, and tragedies, but he is also of interest in the history of science for his *Natural Questions*. This work deals principally with topics in physics, meteorology, and astronomy, where his sources are principally Aristotle and Theophrastus. He wrote of the low state to which general knowledge of astronomy had fallen since the days when the ancient Greeks "counted the stars and named every one," noting that "there are many nations at the present hour who merely know the face of the sky and do not understand why the moon is obscured in an eclipse." But he was hopeful that "the day will come when the progress of research through the long ages will reveal to sight the mysteries of nature that are now concealed."

During the early medieval period the attitude of Christian scholars was that the study of science is not necessary, for in order to save one's soul it is enough to believe in God, as Saint Augustine of Hippo (354–430) wrote in his *Enchiridion:* "It is enough for Christians to believe that the only cause of created things, whether heavenly or earthly, whether visible or invisible, is the goodness of the Creator, the one true God, and that nothing exists but Himself that does not derive its existence from Him."

The Greek philosopher Plotinus (205–270) is believed to have been born in Egypt, where he studied at Alexandria before moving to Rome at the age of forty. His works cover the whole range of philosophy, including cosmology and physics. They embody a synthesis of Platonic, Pythagorean, Aristotelian, and Stoic thought that came to be known as Neoplatonism, the dominant philosophy in the Graeco-Roman world through the rest of antiquity and on into the medieval era.

The best known of his disciples was Porphyry (232/3–c. 305), who was of Syrian origin and had studied in Athens before going on to Rome to study under Plotinus. His numerous and varied writings include a biography of Plotinus and treatises on philosophy, science, religion, and philology, as well as commentaries on works of Plato, Aristotle, and Theophrastus. Though not an original thinker, Porphyry was an important source of information on earlier Greek thought for medieval Latin scholars in western Europe.

The Roman writer Chalcidius, who flourished in the fourth century, is noted for his Latin translation of Plato's *Timaeus*, as well as for his commentary on that work. These were the only sources of knowledge about Plato's cosmology available in Europe during the early medieval era. Chalcidius was also influenced by the ideas of Aristotle and transmitted them to medieval Europe as well, though in somewhat modified form. Two of the most important ideas of Aristotelian science that were thus perpetuated were the concept of the four elements and the astronomical theory of the homocentric spheres, along with the notion of the dichotomy between the terrestrial and celestial regions. Chalcidius refers to the astronomical theories of Heraclides Ponticus, who is also mentioned by the Roman Neoplatonists Macrobius and Martianus Capella.

Macrobius, who may have been from North Africa, flourished in the early fifth century. His most important extant work is his *Commentarii*, a commentary on Cicero's *Dream of Scipio*. Here he used passages of Cicero's work, along with a lost commentary on Plato's *Timaeus*, to construct a treatise on Neoplatonic philosophy. William H. Stahl referred to the *Commentarii* as "the most satisfactory and widely read Latin compendium on Neoplatonism that existed in the Middle Ages." Besides his mention of the theories of Heraclides, Macrobius also wrote of the number mysticism of the Pythagoreans, and said that several Platonists believed that the interplanetary distances were such as to produce harmonious relations, the famous "harmony of the spheres."

Martianus Capella (c. 365–440) was born in Carthage and appears to have been a secondary school teacher or rhetorician. He is the author of an allegorical work entitled *The Nuptials of Mercury and Philology*, in which a series of seven bridesmaids present a digest of each of the seven arts. Book 8 of this work is an introduction to astronomy, in which he states that Mercury and Venus orbit the sun, a theory that he attributes to Heraclides Ponticus, though probably mistakenly. His work was very popular in the early medieval era in Europe, when a number of commentaries were written on it.

Thus a faint glimmer of the light of classical learning remained visible in the darkness of the early Middle Ages, all that survived from the burning of the great Library of Alexandria. But despite the low level of learning in the early medieval era, the first steps were being taken on the road toward an intellectual revival. Progress would be glacially slow at first, for western Europe had only a few fragments of classical learning to start with and most of the population was illiterate. But in time, as we will see, the development of the monastic movement not only increased literacy, but it produced the first European scientists, who would lay the foundations for the emergence of modern science, a process that would begin far from the former centers of ancient Graeco-Roman culture.

2

Educating Europe

*B*Y THE BEGINNING OF THE SIXTH CENTURY ROMAN CIVILIZATION had virtually come to an end in western Europe, except for some fragments of classical learning preserved by a few scholars in monasteries, most that of Monte Cassino, south of Rome, founded by Saint Benedict (d. c. 550). But from these meager fragments the intellectual revival would begin, as we shall see, in what had been the outermost fringes of the Roman Empire in the West, where a spark of classical learning had been kept alive through the Dark Ages and would come alight again to begin the reeducation of Europe.

The most important figure in the transmission of ancient Greek knowledge to early medieval Europe is Anicius Manlius Severinus Boethius (c. 480–525). Boethius, who was from an aristocratic Roman family, held high office under the Ostrogoth king Theodoric, who had him imprisoned and executed. His best-known work is his *Consolation of Philosophy*, written while he was in prison. I first read this work in the Great Books Program of St. John's College of Annapolis, Maryland, whose catalogue had been given to me by the Catholic chaplain of the troopship on which I crossed the Pacific in the last year of World War II. I read through the entire curriculum on my own during the year after I was discharged from the U.S. Navy, one month before my twentieth birthday. It was probably my reading of the *Consolation of Philosophy*, the last great work of Roman scholarship, which inspired me to resume my education at Iona College the following year.

The other works of Boethius fall into two categories, his translations from Greek into Latin of Aristotle's logical works, and his own writings on logic, theology, music, geometry, and arithmetic. His writings were influential in the transmission to medieval Europe of the basic parts of Aristotle's logic and of elementary arithmetic.

The grand plan envisioned by Boethius was to transmit, in Latin, the intellectual achievements of the ancient Greeks. As he wrote in one of his commentaries: "I shall transmit and comment upon as many works by Aristotle and Plato as I can get hold of, and I will try to show that their philosophies agree." He also noted, in the introduction to his *Arithmetic*, a handbook for each of the disciplines of the quadrivium: arithmetic, geometry, astronomy, and musical theory. The extant writings of Boethius also include translations and/or commentaries on at least five logical works by Aristotle, as well as translations of Euclid's *Elements* and Porphyry's *Introduction to Aristotle's Logic*.

Cassiodorus (c. 490–580) was another Roman who held high office under the Ostrogoths. In his *Introduction to Divine and Human Readings* he urges the monks of the monastery at Monte Cassino to copy faithfully the classics of ancient scholarship preserved in their libraries as a cultural heritage. He listed some of the important works of science that he thought should be preserved, and in doing so he described and thus transmitted the basic Aristotelian classification of the sciences. This classification divides philosophy into theoretical and practical areas. The theoretical areas are metaphysics, physics, and mathematics, where the latter is further subdivided into arithmetic, music, geometry, and astronomy, while the practical areas are ethics, economics, and politics. The mathematical sections of the book are briefer and more elementary, mostly dealing with definitions. There is also a section on medicine, in which Cassiodorus gives advice on the use of medicinal herbs, urging the monks to read the works of Hippocrates, Galen, and Dioscorides.

Learn, therefore, the nature of herbs, and study diligently the way to combine various species ... and if you are not able to read Greek, read above all the *Herbarium* of Dioscorides, who described and drew the

Boethius teaching his students, from a manuscript of the
Consolation of Philosophy, *1385*

herbs of the field with wonderful exactness. After this read translations
of Hippocrates and Galen, especially the *Thereapeutics* ... and Hippocrates' *De Herbis et Curis*, and diverse other books written on the art
of medicine, which for God's help I have been able to provide for you
in our library.

Isidore of Seville (c. 560–636), a Visigothic bishop, wrote the first European encyclopedia, the *Etymologies*, which incorporated compilations of
all the Roman authors to whom he had access. The sections on science deal
with mathematics, astronomy, human anatomy, zoology, geography, meteorology. geology, mineralogy, botany, and agriculture. He also wrote a book
called *De rerum natura* (*On the Nature of Things*), taking his title from Lucretius' didactic poem on the atomic theory. Isidore's works contain nothing
original, but they are a very careful compendium of the scientific lore and
other general knowledge available in his time. Despite the very low scientific

level of this work, it achieved wide popularity in early medieval Europe as a source of knowledge of all kinds from astronomy to medicine.

The lifetime of the Greek geographer Strabo (63 BC–c. AD 25) extended from the end of the Ptolemaic period through the first half-century of the Roman imperial era. He was born in Amaseia in the Pontus, and in his youth he studied first in Nysa in Asia Minor, then in Alexandria, and later in Rome. He lived for a long time in Alexandria, where he would have studied the works of Eratosthenes and other Greek geographers whom he mentions. His earlier historical work is lost, but his more important *Geography* in seventeen books has survived. Strabo followed the tradition of Eratosthenes in geography but added encyclopedic descriptions "of things on land and sea, animals, plants, fruits, and everything else to be seen in various regions."

Strabo, writing of Britain early in the first century AD, compared its inhabitants to the Celti, or Celts, of Ireland and continental Europe: "The men of Britain are taller than the Celti, and not so yellow-haired, although their bodies are of looser build.... Their habits are in part like those of the Celti, but in part more simple and barbaric—so much so, although well supplied with milk, they make no cheese; and they have no experience in gardening or other agricultural pursuits."

He went on to describe Ireland, which he called Ierne: "Concerning this island I have nothing certain to tell, except that its inhabitants are more savage than the Britons, since they are man-eaters as well as herb-eaters, and since further, they count it an honorable thing when their fathers die, to devour them, and openly to have intercourse, not only with the other women, but also with their mothers and sisters." I hesitate to point this passage out to my Irish relatives and friends, though given the Irish sense of humor, I'm sure they would find it laughable.

Strabo is indulging in his usual practice of describing the peoples at the far limits of the known world as barbarous or even fabulous, though at least with the Britons and Irish he admits that he has no direct knowledge of them. But by ancient Greek standards they would certainly have been considered uncivilized, if for no other reason than that they were uneducated. This would change only when Christianity came to the British Isles, creating a literate elite and the monasteries that perpetuated learning through schools, libraries, and the production of books.

The first Irish monasteries were probably founded by monks from Britain and Gaul fleeing the Anglo-Saxon and Germanic invasions of the fifth and sixth centuries. A few monks from the Greek world also came to both Ireland and Britain, bringing with them their language and certain texts that were not found elsewhere in Europe, and may have founded some monasteries as well.

Saint Columba (521–597) was born in what is now county Donegal in Ireland, a member of the prominent O'Neil clan who claimed to be descendents of the pagan High King Niall of the Nine Hostages (r. c. 368–395). The nine hostages included one from each of the five provinces of Ireland—Ulster, Connacht, Leinster, Munster, and Meath—and one each from Scotland, the Saxons, the Britons, and the Franks. Tradition has it that Columba was Niall's great-great-grandson. I have learned that my father's family, who were said to be the hereditary keepers of the Stone of Doon, on which the kings of Ireland were crowned, are also descendants of Niall, a claim that can be made by a significant percentage of the population of modern Ireland.

Columba founded monasteries in Ireland in Derry and Durrow, and then in 561, after being defeated at the battle of Cul Dreimhne, he went into exile in Scotland along with twelve companions. He was granted land off the west coast of Scotland on the island of Iona, which became the center of his evangelizing mission to the Picts.

Columba's younger contemporary, Saint Columbanus (c. 540–615), was a monk from Bangor in Ulster, who left Ireland with a group of companions for Gaul, where he founded a number of monasteries centered in Luxeuil on the Vosges. The last of his foundations was at Bobbio in Lombardy, which would become the greatest of the Irish monasteries on the continent.

Saint Benedict Biscop (c. 628–690) was the founding abbot of the monasteries in Wearmouth and Jarrow on Tyne in Northumbria, for which he made five trips to Rome to bring back building materials, artisans, and books to stock their libraries. The library at Wearmouth became famous throughout medieval Europe, its most prized possession being the *Codex Amiatinus*, the earliest manuscript of the complete Bible in the Latin Vulgate version, the translation done by Saint Jerome.

On his third trip to Rome Benedict returned with Theodore of Tarsus (602–690), a Greek from Asia Minor who had studied in Antioch and Constantinople before joining a monastery in Rome, where in 668 Pope Vitalianus (r. 757–768) appointed him as archbishop of Canterbury. Benedict also brought back the monk Hadrian the African, a Greek-speaking Berber from Tunisia who had twice turned down appointments to the archbishopric of Canterbury.

Theodore and Hadrian founded a monastic school in Canterbury that began what has been called the golden age of Anglo-Saxon scholarship. According to the Venerable Bede, Theodore and Hadrian "attracted a large number of students, into whose minds they poured the waters of wholesome knowledge day by day. In addition to instructing them in the holy Scriptures, they also taught their pupils poetry, astronomy, and the calculation of the church calendar.... Never have there been such happy times as these since the English settled Britain."

Theodore and Hadrian taught Greek and Latin as well as arithmetic and physical science. Bede says that in his own time, early in the eighth century, there were students of Theodore and Hadrian who spoke Greek and Latin as fluently as their own language. Students from Wearmouth and Jarrow went on to become abbots of Benedictine monasteries elsewhere in England, spreading the knowledge they had gained in Canterbury.

The Benedictine Rule, established by Saint Benedict himself, regulated every aspect of a monk's life, obliging him to devote most of his waking hours in worship, meditation, and manual labor. Worship included reading the Bible and prayer books, which meant that the monks, many of whom entered the monastery as children, had to be taught how to read and write. Many of the monasteries had a library and a scriptorium, a room where books needed by the monks were written out by copyists.

The Venerable Bede (674–735), writing in 731, says that at the age of seven he was entrusted to the care of Benedict Biscop at Wearmouth, where he began the studies that he would continue for the rest of his life there and at Yarrow. As Bede wrote of his enrollment at Yarrow: "From that time, spending all the days of my life in residence at this monastery, I devoted my-self wholly to Scriptural meditations. And while observing the regular dis-cipline and the daily round of singing in the church, I have always taken delight in learning, or teaching, or writing."

Like many of his contemporaries, Bede was not what we may typically call a chaste monk. A passage in one of Bede's works, the *Commentary on the Seven Catholic Epistles*, gives the impression that he was married. He wrote, "Prayers are hindered by the conjugal duty because as often as I per-form what is due to my wife I am not able to pray." He made a similar state-ment in his *Commentary on Luke*: "Formerly I possessed a wife in the lustful passion of desire and now I possess her in honourable sanctification and true love of Christ."

Bede's best-known work is the *Historia ecclestiastica gentis Anglorus* (*The Ecclesiastical History of the English People*), completed in about 731. This covers the period from 55 BC, when Julius Caesar added Britain to the Roman Empire, up to the time when the work was completed. After a brief account of the history of Christianity in Roman Britain, Bede tells the story of the mission of Saint Augustine of Canterbury, which converted the Anglo-Saxons. Augustine was the prior of a Benedictine monastery in Rome in 595, when Pope Gregory the Great sent him to Britain to convert the pagan Anglo-Saxon king Aethelbert, ruler of the kingdom of Kent. Au-gustine and his companions landed in Kent in 597 and made their way to Aethlelbert's capital in Canterbury, converting the king and many of his subjects. Augustine was appointed as the first archbishop of Canterbury, and soon after his death in 604 he was revered as a saint.

Bede goes on to tell of the further progress of Christianity in Britain, culminating with the Synod of Whitby in 664, when the Anglo-Saxon cler-ics decided to affiliate themselves with the Roman church, thus broadening Catholicism's reach and connecting Britain to the European mainland. He gives central importance to the conversion of Edwin, king of Northumbria, in 627, for once he and other Anglo-Saxon kings became Christians, they bestowed patronage on Irish and Benedictine monasteries and their schools.

Bede's *Ecclesiastical History* tells of miraculous cures made by the relics of saints, including that of a monk who was cured of palsy at the tomb of Saint Cuthbert:

> Falling prostrate at the corpse of the man of God, he prayed with godly earnestness that through his help the Lord would become merciful unto him: and as he was at his prayers, ... he felt (as he afterwards was wont to tell) like as a great broad hand had touched his head in that place where the grief was, and with that the same touching passed all along that part of his body, which had been sore vexed with sickness, down to his feet, and little by little the pain passed away and health followed thereon.

Bede wrote several *opera didascalica*, textbooks designed for vocational courses in monastic schools such as *notae* (scribal work), grammar (literary science), and *computus* (the art and science of telling time). The study of *computus* was necessary in the first place for regulating the daily, weekly, monthly, and yearly activities and festivals in monastic communities, and subsequently it was applied to the needs of farmers, mariners, geographers, physicians, historians and other scholars, tradesmen, and craftsmen. The dating of Easter, for example, was the subject of the Easter *computus*. By the time it was fully developed the *computus* comprised most of the content of the mathematical quadrivium: arithmetic, geometry, astronomy, and musical theory. Bede's *computus* texts included a theoretical section, with explanations of rules and formulas; a practical section that contained a chronicle of world history listing eclipses, earthquakes, and other natural disasters as well as human catastrophes; and appendices including various calendars, chronological tables, and formularies involving the reckoning of time.

Bede's scientific works include a text entitled *De temporibus* (*On Time*), published in 703, which gives an introduction to the principles of the Easter

The Venerable Bede, from a medieval manuscript

computus as well as a chronology of world history. His *De temporum ratione* (*On the Reckoning of Time*), published in 725, is a rewriting of *De temporibus*, which he lengthened tenfold. As he wrote of his earlier work: "When I tried to present and explain it to some brothers, they said it was far more condensed than they wanted, especially the calculation of Easter, which seemed most useful; and so they persuaded me to write somewhat more at length about the nature, course, and end of time."

De temporum ratione presented the traditional view of the geocentric cosmos, explaining how the spherical earth influenced the changing length of daylight during the year, of how the appearance of the sickle moon at twilight was due to the motion of the sun and moon, and of how to relate the occurrence of the tides at a given place to phases of the moon. He discussed spring and neap tides and suggested that tidal action followed a nineteen-year cycle related to the motions of the sun and moon. He was also the first to state the important principle behind what is now known as the establishment of a port, which says that the times of high and low tides at places along a given coast, such as that of Northumbria, depend on their relative location. He wrote: "Those who live on the same shore as we do, but to the north, see the ebb and flow of the tides well before us.

In every place the moon always keeps the rule of association, which she has accepted once and for all." Besides these works on astronomical timekeeping, Bede also wrote a book called *De rerum natura*, based largely on the *Etymologies* of Isidore of Seville and Pliny's *Natural History*, which was not available to Isidore. Although derivative, *De rerum natura* is greatly superior to the *Etymologies* and the *Natural History*, partly because he had access to the works of both Isidore and Pliny, and also because Bede's approach is much more critical than theirs.

Bede's *De rerum natura* presents his views on cosmology, based on the Aristotelian model in which the terrestrial region of the central earth is surrounded by the nested spheres of the celestial region. Whereas Isidore thought that the earth was wheel-shaped, Bede held that it is an immobile sphere divided into five zones, with a central tropic zone, above and below, which are temperate and arctic regions, with only the northern temperate zone inhabited. The terrestrial region was made up of earth at the center surrounded by water, air, and fire. Surrounding the earth are the seven nested spheres of the celestial region: air, aether, Olympus, fiery space, the firmament with the heavenly bodies, the heaven of the angels, and, above all, the heaven of the Trinity. Using Pliny as his source, Bede gave a detailed description of the apparent daily and yearly motion of the sun, moon, stars, and planets, which he had moving within the celestial sphere in a system of epicycles. He also gave a clear explanation of the phases of the moon and of solar and lunar eclipses.

The dating of Easter was one of the questions settled at the Synod of Whitby, where it was decided that the British church would follow Rome in celebrating Easter on the first Sunday after the first full moon following the spring equinox. The problem was to calculate this date in advance from the solar and lunar cycles that determine the times of an equinox and a full moon, respectively.

Bede solved this problem in a truly scientific way. Since *De tempore ratiorum* was focused on the Easter *computus*, Bede gave instructions for determining the date of Easter and its related full moon, for computing the progress of the sun and moon through the constellations of the zodiac, and for other calculations related to the calendar. He also made a new calculation of the age of the world since the Creation, which he dated to 3,952 years before the beginning of the Christian era.

The time between one full moon and the next is either 29 or 30 days, averaging close to 29.5 days over the course of an astronomical year, defined as the time between two successive spring equinoxes. This means that twelve lunar months come to about 354 days, whereas the astronomical year is close to 365.25 days. The Athenian astronomer Meton had around 425 BC found that nineteen astronomical years were almost equal to 235 lunar months. He thus proposed what came to be known as the Metonic cycle, in which a period of nineteen years contains twelve years of twelve months each and seven years of thirteen months, for a total of 235 months. The Metonic cycle was used by the monk Dionysius Exiguus for computations of the date of Easter, a task that had been set for him by Pope John I in 525. Dionysius extended the cycle backward in time to establish the dates of the Resurrection and of the *Incarnationis dominicae tempus* (the time of the Incarnation of the Lord). The latter date came to be used as the starting point of the Christian calendar, in which events were dated from the *annus Domini*, which became the modern AD. Dionysius completed the task set for him by the pope seven years later, when he produced an Easter table for the years we now label AD 532–626, a period of ninety-five years, or five times the nineteen-year Metonic cycle. Bede used this system for dating events throughout the *Historia ecclesiastica*, which was widely used in medieval Europe and helped establish the practice of labeling years of the Christian era as AD. The papacy did not accept the system until 962, when the Saxon Otto I was crowned Holy Roman Emperor. The BC notation originated in 1627, when the Jesuit priest Dionysius Petavius suggested that pre-Christian dates should be labeled *ante Christum*, which in English became "before Christ."

Bede's scientific works were remarkable achievements for their time, and for centuries afterward they were Europe's principal sources of knowledge concerning history, cosmology, chronology, astronomy, natural science, mathematics, and the calendar. His fame as a scholar led a ninth-century Swiss monk to write that "God, the orderer of natures, who raised the Sun from the East on the fourth day of Creation, on the sixth day of the world has made Bede rise from the West as a new Sun to illuminate the whole Earth."

After Bede's death his tradition of learning was kept alive by his student and disciple Egbert, archbishop of York. The cathedral school that Egbert founded in York rivaled the famous monastic schools in Wearmouth and Jarrow, renowned not only in religious studies but in the seven liberal arts of the trivium and quadrivium as well as in literature and science, its library reputed to be the finest in Britain. Generations of successors continued the traditions set by Bede at the York school, which became one of the most respected institutions in Europe.

The most distinguished student at the York school in its early days was Alcuin (c. 735–804), who had been enrolled there as a young boy by Egbert. Alcuin became a teacher at the school in the middle 750s under Egbert, and in 778 he was appointed headmaster and chief librarian. He was sent to Rome in 781 to petition the pope for official confirmation of York's status as an archbishopric, and also to confirm the election of his close friend Eanbld as archbishop. On his way home he passed through Parma, where he met Charlemagne, "King of the Franks," who had in 768 inherited a realm now known as the Carolingian Empire, which included portions of what is now Germany and most of present-day France, Belgium, and Holland, to which he eventually added more German territory, all of Switzerland, part of Austria, and half of Italy. Charlemagne asked Alcuin to join the court at Aachen as his principal adviser on ecclesiastical and educational matters. Alcuin was reluctant to leave his beloved school in York, but he subsequently accepted the offer because, as he later wrote, he felt that God was calling him to the service of Charlemagne.

He arrived in Aachen in 782, bringing with him from York three of his assistants, Pytell, Sigewulf, and Joseph. He was appointed head of the Palace School, which had been founded by the earlier Merovingian kings to educate the royal children, principally in courtly ways and manners. Charlemagne wanted Alcuin to broaden the education of his children and other members of his court to include the study of religion and the seven liberal arts. Alcuin's students included Charlemagne himself and several members of the Carolingian royal family, as well as young men sent to the court for their education along with clerics at the palace chapel.

Alcuin drew up a standardized curriculum and wrote textbooks and manuals for the Palace School, establishing the ancient trivium and quadrivium as the basis for education. He also introduced the various disciplines of the *computus*, including sufficient mathematics and astronomy for understanding the calendar, particularly the dating of Easter. One of the textbooks attributed to Alcuin is entitled *Propositiones ad acuendos juvenes* (*Problems to Sharpen the Young*). This comprises more than fifty mathematical problems, including four famous riddles involving river crossings, which were revived by some of the most famous mathematicians of the seventeenth century and led to the creation of the branch of modern mathematics called combinatorics. One of Alcuin's river-crossing problems and his solution is given in proposition 18:

Problem: A man had to take a wolf, a goat, and a bunch of cabbages across a river. The only boat he could find only take two of them at a time. But he had been ordered to transfer all of these to the other side in good condition. How could he do this?

Solution: I would take the goat and leave the wolf and the cabbage. Then I would return and take the wolf across. Having put the wolf on the other side, I would take the goat back over. Having left that I would take the cabbage across. I would take the boat across again, and having picked up the goat take it over once more. By this procedure there would be some healthy rowing, but no lacerating catastrophe.

Charlemagne issued an edict in 789, the *admonition generale*, establishing monastic and cathedral schools throughout his realm: "Let schools be established where children can learn to read. Carefully correct the psalms, notation, chant, computus, grammar and the Catholic books on every monastery and diocese, because often some desire to pray to God properly, but they pray badly because of the uncorrected books."

Students at the monastic and cathedral schools included not only clerics but also externs, or laymen, which led to an increase in Latin literacy among Charlemagne's subjects. Alcuin supervised the organization of these schools as well as their curricula and textbooks, which were based on the seven liberal arts. Many of his pupils at the Palace School went on to become bishops and abbots, significantly raising the educational level of the clergy, particularly those who directed and taught at the monastic and cathedral schools.

Charlemagne's establishment of monastic and cathedral schools was basic to his reform of the clergy, with the aim of raising their intellectual as well as moral qualifications so that they could carry out his programs for improving the educational and religious level of his subjects. One of these programs stemmed from his belief that all churches and monasteries in his realm should have a copy of the Vulgate Bible, corrected to eliminate all the scribal errors that had accumulated since Saint Jerome completed his translation in 405. Charlemagne assigned this task to Alcuin, who presented a corrected version of the Vulgate to him in 801. During his work on this project, Alcuin invented a new style of handwriting called Carolingian minuscule, which was far more legible than earlier medieval scripts, since it left spaces between the words and used more extensive punctuation. Once this corrected Vulgate was available, priests could base their teaching and sermons on what the Bible actually said.

The Spanish Visigoth Theodulf (c. 755–821) was a scholar in Charlemagne's court from 766 until about 790, when he was appointed bishop of Orleans. When Theodulf visited Rome in 786, he was impressed by the centers of learning he saw there, and he sent letters to abbots and bishops throughout Charlemagne's realm encouraging them to establish public schools. When he became bishop of Orleans he immediately began to found public schools outside the monastic areas under his control. He emphasized the importance of the seven liberal arts in education, including pagan sources, his own particular favorites being Virgil and Ovid. Theodulf believed that everyone had the right to an education, both clerics and laypeople, women as well as men, rich or poor. He was a pioneer in public education and was one of those responsible for the founding of public schools built outside monastic lands or adjacent to churches.

Charlemagne's educational reforms gave rise to a cultural and intellectual revival known as the Carolingian Renaissance, whose effects can be seen in art, architecture, music, literature, and scholarship, both religious and secular, setting the stage for the beginning of European science.

Alcuin returned to England in 790 and remained there until mid-792, when he rejoined the court of Charlemagne as his advisor on religious affairs. At the Council of Frankfurt in 794, Alcuin spoke in support of orthodox doctrine against the views of Felix of Urgel, who was condemned as a heretic. Alcuin retired from his court duties in 796 and was appointed abbot of Saint Martin's monastery in Tours where he died in 804, four years after Charlemagne was crowned as the first Holy Roman Emperor.

The Carolingian Renaissance continued after Charlemagne's death in 814, during the reigns of his son Louis the Pious (r. 814–840) and his grandson Charles the Bald (r. 843–877). Around 845 Charles brought to his court the Irish scholar John Scotus Eriguena (c. 815–c. 877), appointing him head of the Palace School ("Scotus" was the Latin equivalent of "Irish" or "Gaelic"). John is believed to have remained in France for the rest of his days. Something of the character of John and King Charles can be gleaned from an anecdote related by William of Malmesbury. When John was dining with Charles, the king asked him in jest, *"Quid distat inter sottum et Scottum?"* (What separates a sot from an Irishman?), to which he replied, *"Mensa tantum"* (Only a table).

John has been called "the ablest scholar of the ninth century in the Latin West." He was fluent in Greek, which he probably learned at an Irish monastic school, and which he put to good use when he came to the Carolingian court, translating into Latin several Greek patristic works that had been sent to Charles by the Byzantine emperor Michael III. The most important of these was a work by the fifth-century Syrian monk known as Dionysius the Pseudo-Areopagite, since he has been confused with the Aereopagite (Saint Dionysius, who heard Saint Paul's sermon on the Unknown God on the Aereopagus Hill in Athens). Using these and other sources, including Saint Augustine, John then wrote his *Periphyseon*, also known as the *Divisions of Nature*, which has been described as a "highly Neoplatonist system ... based on that school's view of a cosmic process in which all things proceed from the deity and return to Him." The *Periphyseon*, which was later condemned as heretical, had little influence on John's contemporaries, but it led to considerable discussion later in the Middle Ages.

The *Periphyseon* was lost for centuries until it was rediscovered at Oxford and printed in 1671. Once again it began to receive attention, and remained influential through at least the nineteenth century. Eriguena is mentioned by Schopenhauer as a forerunner of modern philosophy, and a modern scholar has described the *Periphyseon* as a prototype of Hegel's *Phenomenology of Spirit*.

John also wrote a commentary on Martianus Capella's *The Marriage of Philology and Mercury*, which was widely used in monastic and cathedral schools since it summed up the seven liberal arts in a single textbook. John's commentary was used by his disciples, and through them, as well as through his *Periphyseon*, he had a continuing influence on Western thought.

The Carolingian educational reforms produced the first major figure in what would become the new European science: Gerbert d'Aurillac (c. 945–1003), who became Pope Sylvester II (r. 999–1003).

Gerbert was born in humble circumstances in or near Aurillac, in southwestern France, and received his early education at the local Benedictine monastery of Saint Géraud. His precocious brilliance brought him to the attention of Count Borel of Barcelona who took him from the monastery in 967 as his protégé. He continued his studies in Barcelona under the aegis of Borel, who put him under the tutelage of Atto, bishop of Vic. Gerbert concentrated on mathematics, probably studying the works of late Roman writers such as Boethius, Cassiodorus, and Martianus Capella, which would have included some arithmetic, geometry (though without proofs), and astronomy, as well as Pythagorean number theory. He would also have learned the mensuration rules of ancient Roman surveyors as well as the art of computing, with the aid of a special kind of abacus.

While Gerbert was in Barcelona he seems to have come in contact with Islamic manuscripts, though probably only in Latin translation. Gerbert's writings include a letter he sent in May 984 to a certain Lupitus of Barcelona, whom he asks to send a translation he has made of a treatise on astrology, presumably from an Arabic work.

Gerbert himself is credited with a treatise on the astrolabe entitled *De astrolabia*, as well as the first part of a work entitled *De utilitatibus astrolabi*, both of which show Arabic influence. The astrolabe, an astronomical instrument and calculating device, was invented by the ancient Greeks, probably by Hipparchus, and was widely used by Islamic astronomers. *De utilitatibus astrolabi* says the astrolabe can be used "to find the true time of day, whether in summer or wintertime, with no ambiguous uncertainty in the reckoning. Yet this seems more suitable for celebrating the daily office of prayer and to be excessive knowledge for general use. How pleasing and seemly the whole proceeds, when with the greatest reverence at the proper hour under the rule of a just judge, who will not wish the slightest shadow of error, they harmonize completely the service of the Lord."

Gerbert's attested writings also include works on mathematics, one of which is a treatise on the abacus, a calculating device that is believed to have come to the Islamic world from China. He also seems to have used the Hindu-Arabic numbers that were subsequently adapted and became the basis for the ones used in the West today. Thus it would appear that Gerbert was one of the first European scholars to make use of the Graeco-Islamic scientific heritage in developing the new science that was beginning to emerge in western Europe.

Gerbert accompanied Borel and Atto to Rome in 970, when he was introduced to Pope John III. The pope in turn brought him to the attention of Otto I (r. 962–973), the Holy Roman Emperor, who was then residing in Rome. Otto arranged for Gerbert's assignment to Adalboro, archbishop of Rheims, who appointed him master of the cathedral school there. Gerbert reorganized the school with such success that students flocked to it from all parts of the empire, his pupils including the future Robert II of France, and Richer of Saint-Remy, who wrote Gerbert's biography. Gerbert's students are known to have gone on to teach at eight other cathedral schools in northern Europe, where they emphasized the mathematics and astronomy he had learned from Islamic sources in Spain.

Gerbert himself taught the courses of the mathematical quadrivium in Rheims. Richer says that Gerbert, in teaching musical theory, used a monochord, the single-stringed instrument that the ancient Pythagoreans employed to demonstrate the numerical laws of harmony. He also says that Gerbert used three mechanical models of his own design and construction in teaching astronomy. The first was a large celestial sphere made of wood covered with horsehide on which he drew a map of the heavens, showing the horizon, the celestial equator, and the ecliptic, with the constellations marked in different colors. The second was a planetarium, with the earth at its center, and a system of three intersecting metal rings to represent the horizon, the celestial equator, and the ecliptic. Small metal balls were placed along the ecliptic to represent the sun, moon, and five visible planets, each of which Gerbert was able to move separately using an ingenious system of wires. The third was a viewing tube oriented with respect to the celestial equator and ecliptic by means of two metal rings, and with outlines of the major constellations made of wire, so that "if one constellation should be pointed out to anyone, even though he were ignorant of the subject he could locate the others without a teacher."

Gerbert acquired the reputation of being a magician, a legend that seems to have begun in the first half of the twelfth century with William of Malmesbury. William says that Gerbert fled from his monastery to study astrology and the black arts with the Saracens, from whom "he learned what the song and flight of birds portend, to summon ghostly figures from the lower world, and whatever human curiosity has encompassed whether harmful or salutary."

Later Gerbert became a favorite of Otto II (r. 973–983), son and successor of Otto I. In 983 he was appointed abbot of the famous monastery of Bobbio in Lombardy, but the following year he returned to Rheims. He then became the tutor and friend of the youthful Otto III (r. 983–1002), who on his father's death in 983 succeeded to the throne as king of Germany at the age of three. His mother, Theophano, a Byzantine princess, acted as regent until her death in 991, when she was replaced by Otto's grandmother Adelaide and Willigis, archbishop of Mainz. Otto ruled on his own after his fourteenth birthday in 994, and two years later he was crowned in Rome as the Holy Roman Emperor, with his two main advisers being Bishop Adalbert of Prague and Gerbert of Aurillac.

Gerbert became archbishop of Ravenna in 998, and the following year, through Otto's influence, he became pope as Sylvester II, the first Frenchman to hold that office. Gerbert had inspired the young Otto with the vision of reestablishing the Graeco-Roman empire of Constantine the Great in Rome, and so when he was elected to the papacy the name he chose was in emulation of Sylvester I, who had been pope at the time when Constantine the Great had moved the capital to Constantinople. Thus Otto could be seen as the successor to Constantine in the great task of reunifying the empires of East and West in its ancient capital in Rome.

Otto died in 1002 at the age of twenty-two, probably of malaria, and Gerbert passed away the following year, thus ending the impossible dream of uniting Greek East and Latin West at the beginning of the second Christian millennium. But a new era was about to dawn through the efforts of the succession of scholars that began in the monasteries of Ireland and progressed to the throne of Saint Peter with Gerbert, the culmination of the Carolingian Renaissance, educating Europe and opening the way for the emergence of European science. But the new science was to be different from the one that had flourished in the ancient Greek world; because of medieval monks like the Venerable Bede, it would be based not only on philosophy and mathematics, but also on observation of the real world on earth and in the heavens above.

By that time Islamic science was at its peak, having absorbed most of the extant scientific works of the Greeks and going on to produce original works of its own. Thus in the clash of civilizations that began with the Holy War of Mohammed, it appeared for a time as if Islamic culture would come to dominate that of the Christian West. But the revival of Western culture in the monasteries of the West would tip the balance, particularly when European scholarship had advanced to a high enough level so that it could begin to absorb Graeco-Islamic science and philosophy.

3

The Opinions of the Arabs

*M*OST OF THE GREEK PHILOSOPHICAL AND SCIENTIFIC WORKS
that survived made their way through the Hellenized Syriac-
speaking Christians of Mesopotamia to the Islamic world after
762, when the Abbasid caliph al-Mansur founded Baghdad as his new capital.

Baghdad emerged as a great cultural center under al-Mansur
(r. 754–775) and three generations of his successors, particularly al-Mahdi
(r. 775–785), Harun al-Rashid (r. 786–809), and al-Ma'mun (r. 813–833).
According to the historian al-Mas'udi (d. 956), al-Mansur "was the first
caliph to have books translated from foreign languages into Arabic,"
including "books by Aristotle on logic and other subjects, and other ancient
books from classical Greek, Byzantine Greek, Pahlavi, Neopersian, and Syriac." These translations were done at the famous Bayt al-Hikma, or House
of Wisdom, a library founded in Baghdad early in the Abbasid period.

There were several motivations for this translation movement, including the desire of al-Mansur and his immediate successors to elevate the cultural level of their caliphate by gaining access to Greek science and
philosophy. Another reason was to educate the secretaries needed to
administer the Abbasid empire. This is evident in the writings of ibn-
Qutayb (d. 889), whose *Adab al-Katib* (*Education of the Secretaries*) enu-
merates the subjects that state secretaries should learn, including arithmetic,
geometry, and astronomy, as well as practical skills such as surveying, metrology, and civil engineering, for which Greek works in science had to be translated into Arabic to provide the necessary textbooks.

Al-Mahdi commissioned the translation of Aristotle's *Topics* into Arabic from Syriac, a form of Aramaic, into which it had been translated from Greek. Later the work was translated directly from Greek into Arabic. *Topics* taught the art of systematic argumentation, which was vital in discourse between Muslim scholars and those of other faiths and in converting non-believers to Islam, which became state policy under the Abbasids. Aristotle's *Physics* was first translated into Arabic during the reign of Harun al-Rashid, the motivation apparently being its use in theological disputations concerning cosmology between Muslims and Christians, who at that time had a far better grounding in philosophy.

The program of translation continued until the mid-eleventh century, both in the East and in Muslim Spain. By that time most of the important works of Greek science and philosophy were available in Arabic translations, along with commentaries on these works and the original treatises by Islamic scientists that had been produced in the interim. Thus, through their contact with surrounding cultures, scholars writing in Arabic were in a position to take the lead in science and philosophy, absorbing what they had learned from the Greeks and adding to it to begin an Islamic renaissance, whose fruits were eventually passed on to western Europe.

At the beginning of the second millennium the so-called Clash of Civilizations became a convergence of cultures, as Islamic science, then at its peak, began to nourish the newly emergent science of western Europe, not only passing along what it had acquired from the Greeks but also works that its own scholars had produced. The effect was profound, for when the heritage of Graeco-Islamic learning became available to Latin scholars, the development of western European science began to accelerate.

The first of the great Islamic scientists, al-Khwarizmi (fl. 828), is renowned for his treatise *Kitab al-jabr wa'l-muqabalah*, known more simply as *Algebra*, for it was from this work that Europe later learned the branch of mathematics known by that name. In his preface the author writes that the Caliph al-Ma'mun "encouraged me to compose a compendious work on algebra, confining it to the fine and important parts of its calculation, such as people constantly require in cases of inheritance, legacies, partition, law-suits and trade, and in all their dealings with one another, or where surveying, the digging of canals, geometrical computation, and other objects of various sorts and kinds are concerned."

Another of al-Khwarizmi's mathematical works survives only in a unique copy of a Latin translation entitled *De numero Indorum* (*Concerning the Hindu Art of Reckoning*), the original Arabic version having been lost. This work, probably based on an Arabic translation of works by the Indian mathematician Brahmagupta (fl. 628), describes the Hindu numerals that eventually became the digits used in the modern Western world. The new notation came to be known as that of al-Khwarizmi, corrupted to "algorism" or "algorithm," which now means a procedure for solving a mathematical problem in a finite number of steps that often involves repetition of an operation.

Al-Khwarizmi is the author of the earliest extant original work of Islamic astronomy, the *Zij al-Sindhind* (a *zij* is an astronomical handbook with tables). This is a set of planetary tables using earlier Indian and Greek astronomical elements, including the epicycle theory. He and Fadil ibn al-Nawbaht are credited with building the first Islamic observatory, which they founded in Baghdad in around 828, during the reign of al-Ma'mun. Al-Khwarizmi also wrote the first comprehensive Islamic treatise on geography, in which he revised much of Ptolemy's work on this subject, drawing new maps.

Euclid's *Elements* was first translated into Arabic during the reign of Harun al-Rashid by the mathematician al-Hajjaj ibn Matar (fl. c. 786–833). Al-Hajjaj did an improved and abbreviated version of the *Elements* for al-Ma'mun, apparently for use as a school textbook.

Islamic astronomy was dominated by Ptolemy, whose works were translated into Arabic and also disseminated in summaries and commentaries. The earliest Arabic translation of the *Almagest* is by al-Hajjaj ibn Matar in the first half of the ninth century. The most popular compendium of Ptolemaic astronomy was that of al-Farghani (d. after 861), who used the findings of earlier Islamic astronomers to correct the *Almagest*. Habash al-Hasib (d. c. 870) produced a set of astronomical tables in which he introduced the trigonometric functions of the sine, cosine, and tangent, which do not appear in Ptolemy's works.

Islamic science developed apace with the translation movement, which involved philosophers as well as scientists. The beginning of Islamic philosophy is credited to Yaqub ibn Ishaq al-Kindi (c. 795–866), the Latin Alkindes, famous in the West as the "Philosopher of the Arabs." Al-Kindi was from a wealthy Arab family in Kufa, in present-day Iraq, which he left to study in Baghdad. There he worked in the Bayt al-Hikma, enjoying the patronage of al-Ma'mun and his immediate successors.

Al-Kindi, though not a translator himself, benefited from the translation movement to become the first of the Islamic philosopher-scientists, founding the Aristotelian movement in Islam. He was a polymath, his treatises including works in geography, politics, philosophy, cosmology, physics, mathematics, meteorology, music, optics, theology, alchemy, and astrology. He was the first Islamic theorist of music, following in the Pythagorean tradition. His work on optics follows Theon of Alexandria in studying the propagation of light and the formation of shadows, and his theory of the emission and transmission of light is based on that of Euclid. Al-Kindi's ideas on visual perception, which differed from those of Aristotle, together with his studies of the reflection of light, laid the foundations for what became, in the European renaissance, the laws of perspective. His studies of natural science convinced him of the value of rational thought, and as a result he was the first noted Islamic philosopher to be attacked by fundamentalist Muslim clerics.

Hunayn ibn-Ishaq (808–873), known in Latin as Jannitus, was born in al-Hira in southern Iraq, the son of a Nestorian apothecary. He went to Baghdad to study under the Nestorian physician Yuhannah ibn Masawayh (d. 857), personal physician to al-Ma'mun and his successors. He then moved to Baghdad, where he and his students, who included his son Ishaq ibn-Hunayn and his nephew Hubaysh, made meticulous translations from Greek into both Syriac and Arabic. Their translations included the medical works of Hippocrates and Galen, Euclid's *Elements*, and *De materia medica* of Dioscorides, which became the basis for Islamic pharmacology. Ishaq's extant translation of Aristotle's *Physics* is the last and best version of that work in Arabic. His translations included Ptolemy's *Almagest*, while his father, Hunayn, revised the *Tetrabiblos*. Hunayn also revised an earlier translation of Galen by Yahya ibn al-Bitriq (d. 820); these were synopses that contained Plato's *Republic*, *Timaeus*, and *Laws*, the first rendering of the Platonic dialogues into Arabic.

Hunayn was an outstanding physician and wrote two books on medicine, both extant in Arabic, one of them a history of the subject, the other a treatise entitled *On the Properties of Nutrition*, based on Galen and other Greek writers. His other writings include treatises on philosophy, astronomy, mathematics, optics, ophthalmology, meteorology, alchemy, and magic, and he is also credited with establishing the technical vocabulary of Islamic science.

Thabit ibn-Qurra (c. 836–901) was born in the Mesopotamian town of Harran, a center of the ancient Sabean cult, an astral religion in which the sun, moon, and five planets were worshipped as divinities. Harran had preserved Hellenic literary culture, and so educated Sabeans like Thabit were fluent in Greek as well as in Syriac and Arabic.

Thabit translated works from both Syriac and Greek into Arabic, his works including improved editions of Euclid's *Elements* and Ptolemy's *Almagest*. His descendants produced Arabic translations of the writings of Archimedes and Apollonius of Perge, among other works. Thabit's own treatises include works on physics, astronomy, astrology, dynamics, mechanics, optics, and mathematics. He wrote a commentary on Aristotle's *Physics* and an original work entitled *The Nature and Influence of the Stars*, which laid out the ideological foundations of Islamic astrology. He also wrote a comprehensive work on the construction and theory of sundials.

Another prominent figure of the translation movement was Qusta ibn Luqa, a Greek-speaking Christian from Lebanon, who worked in Baghdad as a physician, scientist, and translator until his death in 913. His translations included works of Aristarchus, Hero, and Diophantus. He wrote commentaries on Euclid's *Elements* and *De Materia medica* of Diophantus, as well as original treatises on medicine, astronomy, metrology, and optics. His medical works include a treatise on sexual hygiene and a book on medicine for pilgrims.

Astronomy always held pride of place among the sciences in Islam, and Arabic astronomers often waxed eloquent in extolling the utility and godliness of their field. Muhammed ibn Jabir al-Battani (858–929) begins his *Zij al-Sabi* by citing a verse of the Kuran in praise of astronomy. "He it is who appointed the sun a splendor and the moon a light, and measured for her stages, that ye might know the number of the years, and the reckoning."

Al-Battani, the Latin Albategnius, was a Sabean from Harran who had a private observatory in the Syrian town of al-Raqqa. His *Zij al Sabi*, known in its Latin translation as *De scientia stellarum* (*On the Science of Stars*), was used in Europe down to the end of the eighteenth century. In the preface to his *Zij*, al-Battani writes that the errors he found in earlier astronomical treatises had led him to improve the Ptolemaic model with new theories and observations, just as Ptolemy had done with the work of Hipparchus and other predecessors. Ptolemy had measured the rate of precession to be 1° in one hundred years, while al-Battani found it to be 1° in sixty-six years, whereas the correct value is 1° in seventy-two years. Al-Battani's astronomical writings were translated into Latin and were used by astronomers up to the seventeenth century.

Many Arabic astronomers doubled as astrologers, as did some in western Europe, most notably Kepler. The first Arabic philosopher to attack astrology was Abu Nasr al-Farabi (c. 870–950), known in the West as Alfarabius, whose scientific works include commentaries on Euclid's *Elements* and Ptolemy's *Almagest*.

Astrology was also attacked by Sa'di of Shiraz, the famous thirteenth-century Persian poet. One of his tales tells of an astrologer who returned home unexpectedly and found his wife in bed with a stranger. When he raised a fuss about this, a stranger mocked him by saying, "What can you know of the celestial sphere when you cannot tell who is in your own house."

The first great writer in Islamic medicine is Abu Bakr Muhammed ibn Zakariya al-Razi (c. 865–c. 930), the Latin Rhazes, who was born in the Persian city of Rai. He was famous as a physician in both the East and the West, where he was known as the Arabic Galen. He studied in Rai and became the director of the hospital there. He later headed the hospital in Baghdad, where students came from afar to study with him. He is credited with 232 works, of which most are lost, including all of his philosophical treatises. The most important of his surviving medical works is *al-Hawi*, known in its Latin translation as *Continens*, the longest extant Arabic work on medicine. His treatise on smallpox and measles, known in Latin as *De peste*, was translated into English and other Western languages and published in forty editions between the fifteenth century and the nineteenth.

Ibn Sina (980–1037) was born and educated near Bukhara (in present-day Uzbekistan), and later he lived in the Persian towns of Rai and Hamadan, where he died. He is credited with some 270 works; the best known are the *Canon of Medicine* and the *Book of Healing*, which also contain chapters on logic, ethics, mathematics, physics, optics, chemistry, biology, botany, geology, mineralogy, meteorology, and seismology. He also wrote on the classification of the sciences, ranking philosophy as "queen of the sciences." His medical writings, along with those of al-Razi, were translated into Latin and used as basic texts in Europe's medical schools until the seventeenth century. His *Canon of Medicine* was far ahead of its times in dealing with such matters as cancer treatment, the influence of the environment, the beneficial effects of physical exercise, and the need for psychotherapy, where he recognized the connection between emotional and physical states, including the heartache of unrequited love.

Ibn Sina was the first Muslim scientist to revive the impetus theory of John Philoponus, an attempt to explain why a projectile continues to move after it is fired. He described this impetus as a "borrowed power" given to the projectile by the source of motion, "just as heat is given to water by a fire."

Ibn Sina had immense influence on the subsequent development of science, both in the Islamic world and in Latin Europe, where, as Avicenna, he was known as the Prince of Physicians. His ideas, which combined Platonic and Aristotelian concepts, had a profound effect on Western thought in the thirteenth century, when the new European science was being created from Graeco-Arabic sources.

When I am asked about whether Islamic science produced any original works that surpassed those of the ancient Greeks, I point out in particular Abu `Ali al-Hasan ibn al-Haytham (c. 965–c. 1041), known in the West as Alhazen. Ibn al-Haytham was born in Basra, in Iraq, where he studied mathematics and science before going to Egypt during the reign of the Fatimid Caliph al-Hakim. He took up residence near the al-Azhar mosque, teaching and copying Euclid's *Elements* and Ptolemy's *Almagest*, which supported him while he worked on his researches.

Ibn al-Haytham's masterpiece is his *Book of Optics*, which is considered to be one of the most important and influential works ever produced in Islamic science, representing a definite advance beyond what had been achieved by the ancient Greeks in their study of light. The *Optics* was translated into Latin in the late twelfth or early thirteenth century, under the title *Perspectiva*. The *Perspectiva* was the subject of studies and commentaries in Europe up until the seventeenth century, stimulating the study of optics in the Latin West.

The seventh and final book of the *Optics* is devoted to dioptrics, phenomena involving refraction, which also had been studied by Ptolemy. Ibn al-Haytham gives a detailed description of his improved version of Ptolemy's instrument for measuring refraction, which he used to study the bending of light at plane and spherical surfaces with air-water, air-glass, and water-glass interfaces. Ibn al-Haytham's theory introduced a new method, that of resolving the velocity of light into two independent components, one along the normal and the other perpendicular to it, where the first component changed in the refraction while the second remained constant. This approach, called the parallelogram method, was used by a number of European physicists from the thirteenth century onward, in the study of both light and motion. One of his works describes the *camera obscura*, or pinhole camera, the first appearance of the device that eventually led to the development of photography.

Gerbert of Aurillac appears to have been the first Western scholar to write about the all-important astrolabe, the basic astronomical instrument and calculator that Islamic astronomers had inherited from the Greeks, probably invented by Hipparchus.

The first European to follow Gerbert's lead was the German monk Hermann the Lame (1013–1054), who wrote about the astrolabe as well as the chilinder and quadrant, two other astronomical instruments that had been widely used in the Islamic world. The chilinder is a portable sundial designed to give the time for a single latitude, while the quadrant is used to measure the sun's altitude and also give the latitude and time of day. These instruments are described in *De mensure astrolabi* and *De utilitatibus astro-labi*, two works that have been attributed to Hermann, though the first part of the latter work may be by Gerbert d'Aurillac. All three instruments were widely used in the Latin West after they were acquired from Islamic sources.

The first astronomer in western Europe known to have used the astro-labe is Walcher of Malvern, a German monk who had come to England in about 1091. While traveling in Italy Walcher had observed the eclipse of October 30, 1091, and after his return to England he noted that the time of day was considerably different from that recorded by a brother monk at Malvern. At Malvern the following year he observed the eclipse of October 18, using his astrolabe to locate it accurately on the celestial sphere, where in recording its position he uses the Arabic names for three stars as if they were well known to his readers. Using his early observations, Walcher com-piled a set of tables giving the time of new moons from 1036 through 1111, which he thought to be important for use in astrology. The celestial coor-dinates in these tables were worked out by the clumsy methods of Roman fractions, but in a later treatise, written in 1120, he used the system of degrees, minutes, and seconds of arc that Arab astronomers had inherited from the Greeks. Walcher seems to have taken this system from a treatise published in 1115 by Petrus Alphonsus, a Spanish Jew.

The first of the important translators of Graeco-Islamic science from Arabic into Latin is Constantine the African (fl. 1065–1085). Constantine was a Muslim merchant from Carthage in North Africa who visited the

Lombard court in Salerno in southern Italy, where he learned that there was no medical literature available in Latin. He went back to North Africa and studied medicine for three years, after which he returned to Salerno with a collection of medical writings in Arabic, perhaps as early as 1065. A few years later he converted to Christianity and became a monk in the Benedictine abbey at Monte Cassino. There, under the patronage of the famous abbot Desiderius, later Pope Victor III, he spent the rest of his days in making Latin translations or compilations from Arabic medical texts. Constantine is credited with a score of translations, including works of Hippocrates and Galen and the Arabic writer Haly Abbas (c. 925–994), whose *Kitab al-Maliki* he translated as the *Pentegne*, divided into two sections, *theorica* and *practica*.

Constantine's translations were used at the medical school of Salerno, the first in Europe, founded in the mid-eleventh century. These works were introduced into the curriculum under the title of *Ars medicine* or *Articella*, which formed the foundation of a large part of European medical education on into the sixteenth century. Constantine had always emphasized that medicine should be taught as a basic part of natural philosophy, and the *theorica* section of the *Pantegne* provided the basis for this integrated study.

The First Crusade, which began in 1096, led to the establishment of Crusader states in Edessa, Antioch, and Jerusalem, an important factor in opening up Islamic culture to western Europe. One of the earliest examples of this cross-cultural contact is the work of Stephen of Antioch, a translator who flourished in the first half of the twelfth century. According to Matthew of Ferrara, Stephen was a Pisan who went to Syria, probably to the Pisan quarter of Antioch, where his uncle was the Roman Catholic Patriarch.

At Antioch, Stephen learned Arabic and translated the *Kitab al-Maliki* of Haly Abbas into Latin, under the title of *Regalis dispositio*, which he completed in 1127. Stephen says that he did so because he felt that the previous translation of this work by Constantine the African was incomplete and distorted. He also added a prologue to the second part of this work, a list of synonyms in three columns—Arabic, Latin, and Greek—as an aid to help his readers understand the Arabic terms in *De materia medica* of Dioscorides. There he noted that those who have difficulty with the Latin terms can consult experts, "for in Sicily and Salerno, where students of such matters are chiefly to be found, there are both Greeks and men familiar with Arabic."

The leading figure in the early European acquisition of Arabic science was Adelard of Bath (fl. 1116–1142). In the introduction to his *Questiones naturales*, addressed to his nephew, Adelard wrote of his "long period of study abroad," first in France, where he studied at Tours and taught in Laon. He then went on to Salerno, Sicily, Tarsus, Antioch, and probably also to Spain, spending a total of seven years abroad.

Adelard may have learned Arabic in Spain, for his translation of the *Astronomical Tables* of al-Khwarizmi was from the revised version of the Andalusian astronomer Maslama al-Majriti (d. 1071). The *Tables*, comprising thirty-seven introductory chapters and 116 listings of celestial data, provided Christian Europe with its first knowledge of Graeco-Arabic-Indian astronomy and mathematics, including the first tables of the trigonometric sine function to appear in Latin.

Adelard may also have been the author of the first Latin translation of another work by al-Khwarizmi, *De numero Indorum* (*Concerning the Hindu Art of Reckoning*), which describes the Hindu-Arabic numerals that eventually became the digits used in the modern Western world. These numerals, including the all-important zero, may have been introduced in India from Greek sources in Alexandria and were further developed in the Arab world before taking their present form in late medieval Europe. They replaced the unwieldy Roman numerals and were a great stimulus to the development of mathematics in western Europe.

Adelard was probably also the first to give a full Latin translation of Euclid's *Elements*, which he did in three versions. The second of these became very popular, beginning the process that led to Euclid's domination of medieval European mathematics. The first complete English edition appears in Robert Recorde's *The Pathway to Knowledge*, in London in 1551. Recorde realized that Euclid's axioms would be far beyond the mathematical ability of the "simple ignorant" people who would read his book, "for nother is there anie matter more straunge in the English tunge, than this whereof never booke was written before now, in that tunge before now, in that tunge." The first proper English translation, published in London in 1570, was by Sir Henry Billingsley, later lord mayor of London, with a "fruitfull Praeface" by John Dee, who wrote that the book contains "manifolde additions, Scholies, Annotations and Inventions ... gathered out of the most famous and chiefe Mathematicians, both of old time and in our age."

Adelard says that his *Questiones naturalis* was written to explain "something new from my Arab studies." The *Questiones* are seventy-six in number, 1–6 dealing with plants, 7–14 with birds, 15–16 with mankind in general, 17–32 with psychology, 33–47 with the human body, and 48–76 with meteorology and astronomy. Throughout he looks for natural rather than supernatural causes of phenomena, a practice that would be followed by later European writers.

In one particularly interesting passage in this work, Adelard's nephew asks him if it were not "better to attribute all the operations of the universe to God." Adelard replied: "I do not detract from God. Everything that is, is from him and because of him. But [nature] is not confused and without system and so far as human knowledge has progressed it should be given a hearing. Only when it fails utterly should there be a recourse to God."

The *Questiones naturalis* remained popular throughout the rest of the Middle Ages, with three editions appearing before 1500, as well as a Hebrew version. Adelard also wrote works ranging from trigonometry to astrology and from Platonic philosophy to falconry. His last work was a treatise on the astrolabe, in which once more he explained "the opinions of the Arabs," this time concerning astronomy. The treatise describes the workings of the astrolabe and its various applications in celestial measurements, using Arabic terms freely and quoting from Adelard's other works, particularly his translations of Euclid's *Elements* and the planetary tables of al-Khwarizmi.

A woman personifying Geometry, and her students, from a manuscript of the
Elements *of Euclid translated from the Arabic by Adelard of Bath*

Toledo became a center for translation from the Arabic after its recapture from the Moors in 1085 by Alfonso VI, king of Castile and Leon, the first major triumph of the *reconquista*, the Christian reconquest of Andalusia.

The Muslim conquest of the Iberian peninsula began in the spring of 711, when Musa ibn Nusayr, the Arab governor of the Maghreb, or northwest Africa, sent an army across the Strait of Gibraltar under the command of Tariq ibn Ziyad. At that time the Iberian peninsula was ruled by the

Visigoths, whose king, Roderic, was defeated and killed in July 711 by Tariq, who went on to capture Cordoba and Toledo, the Visigoth capital.

Musa followed across the strait with an even larger army and, after taking Seville and other cities and fortresses, he joined Tariq in Toledo. Musa was then recalled to Damascus by the Umayyad caliph, leaving the conquered lands in the hands of his son `Abd al-Aziz, who, in the three years of his governorship (712–715), extended his control over most of the Iberian peninsula, which came to be known to the Arabs as al-Andalus.

The first Abbasid caliph, Abu al-Abbas al-Saffah (r. 749–754), sought to consolidate his power by slaughtering all of the members of the Umayyad family, but one of them, the young prince `Abd al-Rahman, escaped to the Maghreb and then to Spain, where in 756 he established himself in Cordoba, taking the title of *amir*. This was the beginning of the Umayyad dynasty in Spain, which was to rule al-Andalus until 1031. `Abd al-Rahman I (r. 756–788) established Cordoba as his capital, and in the years 784–786 he erected the Great Mosque, which was rebuilt and enlarged by several of his successors. `Abd al-Rahman II (r. 822–852) began the development of science in al-Andalus by sending an agent to the East to buy books, which an anonymous Maghreb chronicler says included astronomical tables as well as works in astronomy, philosophy, medicine, and music.

The Umayyad dynasty in al-Andalus reached its peak under `Abd al-Rahman III (r. 912–961), who in 929 took the title of caliph, emphasizing the independence of al-Andalus from the Abbasid caliphate in the East. This began the golden age of Muslim Cordoba, known to Arab chroniclers as the "the bride of al-Andalus." The golden age continued under `Abd al-Rahman's son and successor, al-Hakem II (r. 961–976), and his grandson Hisham II (976–1009), who was a puppet in the hands of his vizier al-Mansur.

Al-Hakem built one of the greatest libraries in the Islamic world in Cordoba, rivaling those at Baghdad and Cairo. The caliph's library, together with the twenty-seven free schools he founded in his capital, gave Cordoba a reputation for learning that spread throughout Europe, attracting Christian scholars as well as Muslims, not to mention the Sephardic Jews who lived under Islamic rule.

The culmination of Arabic philosophy comes with ibn Rushd, the Latin Averroës (1126–1198), who was from a distinguished family of Cordoban jurists.

He studied theology, law, medicine, and philosophy, including the works of Aristotle, particularly his writings in physics and natural science.

The philosophical writings of ibn Rushd can be divided into two groups, his commentaries on Aristotle and his own treatises on philosophy. He regarded the philosophy of Aristotle as the last word, to the extent that truth can be understood by the human mind. By the beginning of the thirteenth century ibn Rushd was considered to be the outstanding interpreter of Aristotle and his works were translated into Hebrew. By the end of that century nearly half of his commentaries on Aristotle had been translated from Arabic into Latin, so that he came to be known in the West as the Commentator.

Ibn Rushd interpreted the concept of creation in such a way as to deny free will not only to man but even to God himself. According to ibn Rushd, the world had been created by a hierarchy of necessary causes starting with God and descending through the various Intelligences that moved the celestial spheres. He accepted Aristotle's planetary model of the homocentric spheres and rejected Ptolemy's theory of eccentrics and epicycles. He writes of his astronomical researches in his commentary on Aristotle's *Metaphysics*, where he expresses his belief that the prevailing Ptolemaic theory is a mathematical fiction that has no basis in reality.

One of the discoveries made by ibn Rushd in his medical researches was that the retina rather than the lens is the sensitive element in the eye, an idea that was forgotten until it was revived by the anatomist Felix Plater (1536–1614).

After al-Mansur's death in 1002, the caliphate passed in turn to several claimants in the principal cities of al-Andalus, and finally it was abolished altogether in 1031. The fall of the caliphate was followed by a period of sixty years in which al-Andalus was fragmented into a mosaic of petty

Muslim states, allowing the Christian kingdoms of northern Spain to start expanding south. The last remnant of Muslim Spain was the Banu Nasr kingdom of Granada, which hung on until its capture in 1492 by Ferdinand II of Aragon and Isabella of Castile, the Catholic kings, who that year drove the last of the Moors from Spain, also expelling the Jews.

The initiative for the translation movement in Spain seems to have come from Raymond, archbishop of Toledo, (r. 1125–51) as evidenced in the dedications of a contemporary Toledan translator, Domenicus Gundissalinus (c. 1110–c. 1190).

Gundissalinus, archbishop of Segovia, did several translations and adaptations of Arabic philosophy, including works by Al-Kindi, ibn Rushd, and ibn Sina, as well as one by the Jewish physician Isaac Judaeus. The translations attributed to Gundissalinus were probably done by him in collaboration with others who were fluent in Arabic, though only in one work, the *De anima* of ibn Sina, is his name linked with that of a coauthor.

Gundissalinus also wrote five philosophical works on his own, based largely on the books that he had translated as well as on Latin sources. His *De divisione philosophiae*, which incorporates the systems of both Aristotle and Arabic philosophers, is a classification of the sciences transcending the traditional division of studies in the trivium and quadrivium, and it influenced later schemes of classification in western Europe.

Sephardic Jews played an important role in the translation movement, since their languages included Arabic and Latin as well as Hebrew, and several of them wrote original works of enduring importance. The most influential was the mathematician, astronomer, and philosopher Abraham bar Hiyya Ha-Nasi, known in Latin as Savasorda, who was born in Barcelona in 1070 and died in Provence in 1136 or 1145.

Savasorda's most important work is his Hebrew thesis on practical arithmetic, which he and Plato of Tivoli translated into Latin in 1145 as the *Liber embadorum*. This was one of the earliest works on Arabic algebra and trigonometry to be published in Latin Europe, and it contains the first solution of the standard quadratic equation to appear in the West. It was also the earliest to deal with Euclid's *Division of Figures*, which has not survived in Greek and only partially in Arabic. Savasorda's *Encyclopedia* is a compendium of practical reckoning and business arithmetic as well as the theory of numbers and geometric definition. It has been defined as "probably the earliest algorithmic work written in western Europe."

Among the other figures in the translation of Graeco-Arabic science to Latin, the most prolific by far was Gerard of Cremona (1114–1187), whose influence was evident for centuries. The few details that are known of Gerard's life come mostly from a short biography and eulogy written in Cremona after his death. It notes that Gerard completed his education in the schools of the Latins before going to Toledo, which he would have reached by 1144 at the latest, when he would have been thirty years old. The biography goes on to say that it was his love of Ptolemy's *Almagest*, which he knew was not available in Latin, that drew Gerard to Toledo, and "there, seeing the abundance of books in Arabic on every subject ... he learned the Arabic language, in order to be able to translate."

Gerard's translations include Arabic versions of writings by Aristotle, Euclid, Archimedes, Ptolemy, and Galen, as well as works by Thabit ibn-Qurra, al-Kindi, al-Khwarizmi, al-Razi, ibn Sina, and ibn al-Haytham. The subjects covered in these translations include twenty-one works on medicine; seventeen on geometry, mathematics, optics, weights, and dynamics; fourteen on philosophy and logic; twelve on astronomy and astrology; and seven on alchemy, divination, and geomancy, or predicting the future from geographic features.

More of Arabic science passed to the West through Gerard than from any other source. His translations produced a great impact upon the development of European science, particularly in medicine, where students in the Latin West took advantage of the more advanced state of medical studies in medieval Islam, particularly those of ibn Sina. Ibn Sina's *Canon of Medicine*, particularly its encyclopedic detail on the practical side of the healing art, remained unsurpassed up until the beginning of the twentieth century, at least according to the opinion of pharmaco-epidemiology professor John Urquhart. Writing in the *British Medical Journal* in 2006, Urquhart said: "If the year were 1900 and you were marooned and in need of a guide for practical medicine, which book would you want by your side? My choice was Ibn Sina."

Gerard's translations in astronomy, physics, and mathematics were also very influential, particularly those of Archimedes and Euclid, since they represented a scientific approach to the study of nature rather than the philosophical and theological attitude that had been prevalent in the Latin West. Gerard's translation of Ptolemy's *Almagest* was particularly important; as historian Charles Homer Haskins noted, through this work "the fullness of Greek astronomy reached western Europe."

The English scholar Robert of Chester, a younger contemporary of Adelard, collaborated with other translators at several places in southern France and Spain, including Toledo. Robert's solo translations include al-Khwarizmi's *Algebra* (dated Segovia, 1145); a treatise on the astrolabe (London, 1147); a set of astronomical tables for the longitude of London (1149–1150); and a revision, also for the meridian of London, of Adelard's version of the tables of al-Khwarizmi. His treatises on astrolabes and astronomical tables indicate that work on both observational and theoretical astronomy was being done in England at the time.

One of the extant manuscripts of Robert's revisions of al-Khwarizmi's work contains astronomical tables for the longitude of Hereford in England, dated 1178, which have been attributed to Roger of Hereford, who wrote several works on astronomy and astrology in the decade 1170–1180. One of these, a survey entitled *Liber de divisione astronomiae*, begins with the phrase "In the name of God the pious and merciful," the traditional opening of an Islamic treatise, suggesting that this is a translation from the Arabic, though the author is unknown. But interest in the astrolabe and in astronomical tables revised for Hereford indicates that Roger was actively engaged in astronomy himself, the earliest-known astronomer in Latin Europe.

Alfred of Sareshel, another twelfth-century English scholar, dedicated one of his translations to Roger of Hereford. Alfred did translations of several Aristotelian works from Arabic, together with commentaries, and he also translated parts of ibn Rushd's *Kitab al-Shifa*, the sections on geology and alchemy, which he entitled *De mineralibus*. Alfred seems to have learned Arabic in Spain, where he probably did his translation of ibn Rushd, and he also appears to have used Greek sources, particularly in his works on Aristotle, whose natural philosophy and metaphysics he introduced to England.

The most important interface among Greek, Latin, and Arabic culture in the twelfth century was the Norman realm in southern Italy and Sicily, the kingdom of the Two Sicilies. When Count Roger I conquered Palermo in 1091, it had been under Muslim domination for nearly two centuries. He reduced the Muslims to the status of serfs except in Palermo, his capital, where he employed the most talented of them as civil servants, so that Greek, Latin, and Arabic were spoken in the Norman court and used in royal charters and registers. Under his son Roger II (r. 1130–1154), Palermo became a center of culture for both Christians and Muslims, surpassed only by Cordoba and Toledo. Beginning under Roger II, and continuing with his successors, the Sicilian court sponsored numerous translations from both Greek and Arabic into Latin.

Roger II was particularly interested in geography, but he was dissatisfied with existing Greek and Arabic geographical works. Thus in 1138 he wrote to al-Idrisi (1100–1166), the distinguished Muslim geographer and cartographer, who was then living in Cueta, and invited him to visit Palermo, saying, "If you live among the Muslims, their kings will contrive to kill you, but if you stay with me you will be safe." Al-Idrisi accepted the offer and lived in Palermo until Roger's death in 1154, after which he returned to Cueta and passed his remaining days there.

Roger commissioned al-Idrisi to create a large circular relief map of the world in silver, the data for which came from Greek and Arabic sources, principally Ptolemy's *Geography*, as well as travelers and the king's envoys. The map has long since vanished, but its features were probably reproduced in the sectional maps in al-Idrisi's Arabic geographical compendium, *Kitab nuzhat al-mushtaq fi ikhtiraq al-afar*, which has survived. The compendium deals with both physical and descriptive geography, with information on political, economic, and social conditions in the lands around the Mediterranean and in the Middle East, and is thus a veritable encyclopedia of the medieval world. Al-Idrisi's work was a popular textbook in Europe for several centuries and a number of abridgements were done, the first at Rome in 1592. A Latin translation was published at Paris in 1619, and a two-volume French translation was done in 1830–1840, entitled *Géographie d'Edrisi*.

Frederick II of Hohenstauffen (r. 1211–1250), the Holy Roman Emperor and king of the Two Sicilies, was a grandson of the Emperor Frederick I Barbarossa and the Norman king Roger II. Known in his time as *stupor mundi*, "the wonder of the world," he had been raised from age seven to twelve in Palermo, where he grew up speaking Arabic and Sicilian as well as learning Latin and Greek. When he became emperor in 1211, at the age of fourteen, he turned away from his northern dominions to his kingdom of the Two Sicilies, where, like his Norman predecessors, who were known as "baptized sultans," he indulged himself in his harem in the style of an oriental potentate.

Frederick was deeply interested in science and mathematics, and he invited a number of scholars to his brilliant court, most notably John of Palermo, Master Theodorus, and Michael Scot, calling them his "philosophers." He subsidized their scientific writings and translations, which included works of Aristotle on physics and logic, some of which he presented in 1232 to the faculty at Bologna University. The letter that Frederick sent with the gift told of how he had loved learning since his youth, and of how he still took time from affairs of state to read in his library, where numerous manuscripts of all kinds "classified in order, enrich our cupboards."

Frederick's scholarship is evident in his famous book on falconry, *De arte venandi cum avibus*, or *The Art of Hunting with Birds*. This is a scientific work on ornithology as well as a detailed and beautifully illustrated manual of falconry as an art rather than a sport. Frederick acknowledged his debt to Aristotle's *Zoology*, which had been translated by Michael Scot earlier in the twelfth century. But he was critical of some aspects of the work, as he wrote in the preface to his manual: "We have followed Aristotle when it was opportune, but in many cases, especially in that which regards the nature of some birds, he appears to have departed from the truth. That is why we have not always followed the prince of philosophers, because rarely, or never, had he the experience of falconing which we have loved and practiced always." This is the earliest European work critical of Aristotle, a view that is one of Galileo's main claims to fame.

One of those with whom Frederick corresponded was the renowned mathematician Leonardo Fibonacci (c. 1170–after 1240), who had been presented to him when he held court at Pisa around 1225. Fibonacci had at that time just completed his treatise on squared numbers, the *Liber quadratorum*, which he dedicated to Frederick, noting, "I have heard from the Podesta of Pisa that it pleases you from time to time to hear subtle reasoning in Geometry and Arithmetic."

Fibonacci was born in Pisa around 1170. He wrote about his life in the preface to his most famous work, the book on calculations entitled *Liber abbaci*. His father, a secretary of the Republic of Pisa, was around 1192 appointed director of the Pisan trading colony in the Algerian city of Bugia (now Bouge). He was brought to Bugia by his father to be trained in the art of calculating, which he learned to do "with the new Indian numerals," the so-called Hindu-Arabic numbers, which he would introduce to Europe in his *Liber abbaci*. His father also sent him on business trips to Provence, Sicily, Egypt, Syria, and Constantinople, where he met with Latin, Greek, and Arabic mathematicians. Around 1200 he returned to Pisa, where he spent the rest of his days writing the mathematical treatises that made him the greatest mathematician of the Middle Ages.

His extant works are the *Liber abbaci*, first published in 1202 and revised in 1228; the *Practica geometriae* (1220/1221), on applied geometry; a treatise entitled *Flos* (1225), sent to Frederick II in response to mathematical questions that had been put to Fibonacci by John of Palermo at the time of the emperor's visit to Pisa; an undated letter to Master Theodorus, one of the court "philosophers"; and the *Liber quadratorum* (1225). The latter work contains the famous "rabbit problem": "How many pairs of rabbits will be produced in a year, beginning with a single pair, if in every month each pair produces a new pair which become productive from the second month on?" The solution to this problem gave rise to the so-called Fibonacci numbers, a progression in which each number is the sum of the two that precede it:

$$1, 1, 2, 3, 5, 8, 13, 21 \ldots$$

The sequence is a mathematical wonder that continues to fascinate mathematicians. Fibonacci's sources, where they can be traced, include Greek, Roman, Indian, and Arabic works, which he synthesized and, adding to them with his own creative genius, stimulated the beginning of the new European mathematics.

Master Theodorus, who is usually referred to as the Philosopher, was born in Antioch. He served Frederick as secretary, ambassador, astrologer, and translator, from both Greek and Arabic into Latin, and he was also the emperor's chief confectioner. One of his works is a translation of an Arabic work on falconry. He served the emperor until the time of his death around 1250, when Frederick regranted the estate that "the late Theodore our philosopher held so long as he lived."

Theodorus had probably succeeded Michael Scot as court astrologer. Michael was born in the last years of the twelfth century, probably in Scotland. Nothing is known of his university studies, but his references to Paris indicate that he may have studied and lectured there as well as in Bologna, where he did some medical research in 1220–1221. He may have learned Arabic and some Hebrew in Toledo where, around 1217, he translated al-Bitruji's *On the Sphere*, with the help of Abuteus Levita, a Jew who later converted to Christianity. By 1220 he had completed what became the standard Latin versions of Aristotle's *Physica*, *De caelo*, and *De anima* with Averroës's commentaries. He had become a priest by 1224, when Pope Honorius II appointed him as archbishop of Cashel, in Ireland, and obtained benefices for him in England. He declined the appointment as archbishop, saying that he did not speak Irish, and was then given further benefices in England and Scotland by the archbishop of Canterbury.

When Fibonacci completed his revised version of *Liber abbaci* in 1228, he sent it to Michael, who by that time seems to have entered the service of Frederick II as court astrologer and translator. Aside from his translations, Michael's major work was a comprehensive introduction to the sciences, including alchemy, on which, according to historian Lynn Thorndike, he may also have written a separate thesis. Thorndike quoted Michael in describing his procedure for transmuting copper into gold:

"Take the blood of a ruddy man and the blood of a red owl, burning saffron, Roman vitriol, resin well pounded, natural alum, Roman alum, sugared alum, alum of Castile, red tartar, marcasite, golden alum of Tunis, which is red, and salt." These ingredients are to be pounded in a mortar, passed through sieves, treated with the urine of an animal called *taxo*, or with the juice of wild cucumber, then dried, brayed again, and then put into a crucible with the copper.

Michael also wrote a voluminous treatise known in English as *Introduction to Astrology*, a subject that he thought should be studied by all physicians. The treatise covers every aspect of astrology and divination including necromancy, or conjuring up the spirits of the dead to reveal the future or influence the course of coming events, as well as nigromancy, or black magic, dealing with spells cast at night rather than in daylight.

Frederick addressed a long series of extraordinary questions to Michael, who inserted the questionnaire as an addendum to a work entitled *Libers particularis*. Frederick's interest in necromancy is indicated in one of the questions that he asked Michael: "And how is it that the soul of a living man which has passed away to another life than ours cannot be induced to return by first love or even by hate, just as it had been nothing, nor does it seem to care at all for what it has left behind whether it be saved or lost." Michael boasted that he could answer all of the questions asked by the emperor, including his query as to "whether one soul in the next world knows another and whether one can return to this life to speak and show one's self; and how many are the pains of hell."

All of this led to Michael's posthumous fame as a magician, clouding his reputation as a scientist and translator, which is in any event controversial. Roger Bacon referred to Michael as "a notable inquirer into matter, motion, and the course of the constellations," but at the same time he listed him among those translators who "understood neither sciences nor languages, not even Latin," and said that his translations were for the most part done by a Jew named Andrew. Bacon credits Michael with having introduced the natural philosophy of Aristotle to the Latin West, though Michael actually transmitted only three Aristotelian works.

Along with Gerbert d'Aurillac, Michael was said to have sold his soul to the devil in exchange for his knowledge of the black arts and the magic of science. Dante writes of him in canto 20 of the *Inferno*, where he is pointed out in the fourth ditch of the eighth circle of Hell, among the other diviners: "That other, round the loins / So slender of his shape, was Michael Scot / Practised in every slight of magic wile."

European science had by now outgrown the confines of the monasteries where it first developed, spreading into the outer world and interacting with Byzantine and Islamic culture, going beyond the philosophic and mathematical science of the works in the great Library of Alexandria to explore new worlds. What is more, scholarship was no longer restricted to the confines of a monastery, for the level of knowledge in the Latin West was now such that men like Gerard of Cremona and Adelard of Bath now sought out and found new sources of Greco-Islamic science, most notably ibn al-Haytham, whose work on the science of light went well beyond any work of optics produced in the Greek world. And the first step in creating institutions of higher learning had been taken in the foundation of the medical school in Salerno, with its curriculum based on both Greek and Islamic medicine.

These advances had been largely due to the political stability in western Europe at the beginning of the second Christian millennium, which fostered the growth of commerce and increased prosperity. Technological advances such as the improvement and wider use of the waterwheel and the rotation of crops created a much greater agricultural output. This led to a population explosion, and it is estimated that between the years 1000 and 1200 the population of western Europe may have increased by as much as a factor of four. There was an even greater increase in the urban population, which led to more economic opportunity, stimulating intellectual interchange, and the creation of schools, including the first universities of western Europe. The Dark Ages were definitely over; modern Europe was emerging, and with it European science.

Michael Scot is believed to have died in 1236 in Germany, where he had accompanied the emperor Frederick. By that time almost all of Graeco-Islamic science that would be translated from Arabic into Latin was available in western Europe, setting the stage for the next phase of the program, translating works directly from Greek into Latin.

4

A Renaissance Before
the Renaissance

B Y THE TWELFTH CENTURY WESTERN EUROPEAN CULTURE HAD
progressed to the point where scholars were no longer satisfied
with the works of Graeco-Arabic science, but now began to look
for translations directly from the Greek, searching for a deeper insight into
the thought of the great philosophers and scientists of the classical and Hellenistic periods, particularly Aristotle and the great mathematical physicists
and astronomers such as Euclid, Archimedes, and Ptolemy, for their own
researches were taking them deeper into Aristotelianism as well as the use
of mathematics in science.

A few translations from Greek to Latin had been done in Italy during the sixth century, most notably some logical works of Aristotle translated by Boethius. No further translations from Greek into Latin were
done until the twelfth century, when the first cultural interactions began
taking place between the Greek East and the Latin West, principally in
Constantinople, where several Italian city-states had trading concessions,
and in Norman Sicily.

The first important instance of such a cultural interchange occurred
in 1136, during the reign of the Byzantine emperor John II Comnenus,
when the Holy Roman Emperor Lothair sent a mission headed by Anselm,
bishop of Havelberg and later archbishop of Ravenna, to Constantinople
to discuss theological differences between the Roman Catholic and Greek
Orthodox churches. After Anselm arrived in Constantinople he and his
entourage went to the Pisan quarter on the Golden Horn to discuss theology with a group of Greek clerics headed by Nicetas, archbishop of Nico-

media. According to Anselm, "There were present not a few Latins, among them three wise men skilled in the two languages [Latin and Greek] and most learned in letters, mostly James a Venetian, Burgundio a Pisan, and the third, most famous among Greeks and Latins above all others for his knowledge of both literatures, Moses by name, an Italian from the city of Bergamo, and he was chosen by all to be an interpreter for both sides."

The first of these scholars, James of Venice, is known in Latin as Iacobus Veneticus Grecus, which could mean that he was a member of the Greek community of Venice. In any event, he was fluent in both Greek and Latin, as indicated by an entry for the year 1128 in the chronicle of Robert of Torigni, abbot at Mont Saint Michel, who wrote that "James, a clerk of Venice, translated from Greek into Latin certain books of Aristotle and commented upon them, namely the *Topics*, the *Prior* and *Posterior Analytics*, and the [*Sophistici*] *Elenchi*, although there was an older version of these books."

James was the first European scholar in the twelfth century to introduce the works of Aristotle to the Latin West. Besides the works mentioned by Robert of Torigni, James was the first to translate from the Greek Aristotle's *Physica*, *De anima*, *Metaphysica*, and *Parva naturalia*. His commentary on the *Sophistici elenchi* shows that he was aware of Byzantine scholarship on this subject in Constantinople, which was an unrivaled source for the works of Aristotle and other Greek writers. James's translations, together with their revisions, formed the basis for much of Aristotelian studies in Europe up until the sixteenth century.

Burgundio the Pisan traveled frequently from Pisa to Constantinople and to Sicily, another rich source of Greek manuscripts. His translations from the Greek also included the *Aphorisms* of Hippocrates and ten works of Galen, as well as Aristotle's *Meteorology*.

Moses of Bergamo, whom Anselm mentions as the interpreter in the theological disputation of 1136, lived at that time in the Venetian quarter of Constantinople. He wrote in one of his letters that he learned Greek so that he could translate previously unknown manuscripts into Latin. He spent years collecting Greek manuscripts for which he paid a total of three pounds of gold, he says, but they were all destroyed in a fire in 1130.

Translations from Greek to Latin were also done in Sicily during the reign of William I (r. 1154–1166), son and successor of Roger II, who continued his father's patronage of learning. The two principal translators during his reign were Henricus Aristippus and Eugene the Emir, both of them members of the royal administration who have left eulogies of William commemorating him as a philosopher-king who opened his court to the world's leading scholars. Aristippus became archbishop of Catania in 1156 and four years later he was placed in charge of the entire administration of the Sicilian kingdom. He was the first to translate from the Greek two of Plato's dialogues, *Meno* and *Phaedo*, as well as the fourth book of Aristotle's *Meteorology*, works that remained in use until the early Renaissance. Aristippus also served as envoy to the court of Manuel II Comnenus in Constantinople, where the emperor presented him with a beautiful codex of Ptolemy's *Almagest* as a present to King William. The first Latin translation of this manuscript from Greek to Latin was made in Palermo by an anonymous visiting scholar in around 1160. Other works translated from Greek to Latin at the Sicilian court by this scholar include Euclid's *Optica* and *Catoptrica*, the *De motu* of Proclus, and the *Pneumatica* of Hero of Alexandria.

The unknown scholar who translated these works notes that in doing so he received considerable assistance from Eugene the Emir, "a man most learned in Greek and Arabic and not ignorant of Latin." Eugene, who held the Arabic title of emir in the royal administration, was probably a Greek, as evidenced by his surviving poetry.

The Dominican monk William of Moerbeke (b. c. 1220–1235—d. before 1286), in Belgium, was the most prolific of all medieval translators from Greek into Latin. Moerbeke is known to have visited Nicaea in the spring of 1260, when the Byzantines had their capital there until they recaptured Constantinople from the Latins the following year, and he may very well have acquired Greek manuscripts at that time. He took part in the Second Council of Lyons (May–June 1274), whose goal was to bring about a reunion between the Greek and Latin churches, and at a pontifical mass he sang the *Credo* in Greek together with Byzantine clerics.

Thomas Aquinas is said to have suggested to Moerbeke that he complete the translation of Aristotle's works directly from the Greek. Moerbeke says that he took on this task "in spite of the hard work and tediousness which it involves, in order to provide Latin scholars with new material for study."

Moerbeke's Greek translations included the writings of Aristotle, commentaries on Aristotle, and works of Archimedes, Proclus, Hero of Alexandria, Ptolemy, and Galen. The popularity of Moerbeke's work is evidenced by the number of extant copies of his translations, including manuscripts from the thirteenth to fifteenth centuries; printed editions from the fifteenth century onward; and versions in English, French, Spanish, and even modern Greek done from the fourteenth century through the twentieth. His translations led to a better knowledge of the actual Greek texts of several works, and in a few cases they are the only evidence of lost Greek texts, such as that of Hero's *Catoptrica*.

Moerbeke's only original work is a treatise on divination entitled *Geomantia*, which was evidently quite popular, as evidenced by the several extant Latin manuscripts and a French translation done in 1347. The Polish scholar Witelo, in the dedication to Moerbeke in his *Perspectiva*, praises him for his "occult" inquiry into the influence of divine power on humans.

Another Western scholar who visited Constantinople in search of ancient Greek manuscripts was Peter of Abano (1250–c. 1313), who while there found works of Aristotle, Dioscorides, and Galen, among others. His translations from the Greek include a volume of Aristotle's *Problems*, the first Latin translation of this work; *De materia medica* of Dioscorides; and six treatises of Galen.

The most famous of Peter's original works is his *Conciliator differentia cum philosophum et praecipue medicorum*, which he completed in 1303, while he was teaching at the University of Paris. This is an enormous tome in which Peter tries to reconcile the conflicting views of the medical writers and philosophers who had preceded him. The *Conciliator* comprises more than two hundred questions, or "differences," which Peter says he and his colleagues had been debating for the past decade. The first and eighteenth questions, for example, concern the differences of opinion about whether the heart is the center of the human nervous system, as Aristotle holds, or whether it is the brain, as Peter says. His conclusion concerning the first question is that "the regulative power of the body resides in the brain," and to the eighteenth that "the brain is the seat of sensation and emotion." Question 67 asks, "Is life possible below the equator?" It seems that this question occurred to Peter when he met Marco Polo in Constantinople in 1295, after the Venetian's celebrated journey to the Far East.

Another well-known work by Peter is his *Lucidator dubitabilium astronomiae*, in which he discusses disputed doctrines in astronomy and astrology. Here he suggests that the stars are not fixed in the outermost celestial sphere, as Aristotle has it in his model of the cosmos, but that they are moving freely in space, an entirely new idea that would become part of modern cosmology. A number of passages in this work indicate that Peter associated spirits and intelligences with the celestial bodies, one of which he describes as "perpetual and incorruptible, leading through all eternity a life most sufficient unto itself, nor ever growing old."

Peter's writings on astrology and other occult sciences gave him the reputation of being a magician. French librarian and scholar Gabriel Naude, writing in 1625, calls Peter "a man who appeared as a prodigy and miracle in his age ... he was the greatest magician of his age and learned the seven liberal arts from seven familiar spirits whom he held captive in a crystal." He goes on to say that Peter had in later life abandoned "the idle curiosity of his youth to devote himself wholly to philosophy, medicine and astrology."

By the end of the twelfth century European science was on the rise, stimulated by the enormous influx of Graeco-Islamic works translated into Latin from Arabic as well as other works translated directly from Greek. This is evidenced by the opening of the medical school at Salerno and the study of ibn Sina; the work in mathematical and observational astronomy by Roger of Hereford; the beginning of studies of both physical and descriptive geography stimulated by the work of al-Idrisi; the acquisition of the Hindu-Arabic numerals by Leonardo Fibonacci and his researches in number theory; and the critical attitude toward Aristotle shown by Frederick II in his treatise on falconry.

By the first quarter of the thirteenth century virtually all of the scientific works of Aristotle had been translated into Latin, from Greek as well as Arabic, along with the Aristotelian commentaries of Averroës (ibn Rushd). The translations included other works by both Greek and Islamic scientists on optics, catoptrics (reflection), geometry, astronomy, astrology, zoology, botany, medicine, pharmacology, psychology, and mechanics. This body of knowledge became part of the curriculum at the first universities that began to emerge in the late twelfth and early thirteenth centuries, supplanting the monastic and cathedral schools of the earlier medieval era.

All of this led to a cultural revival that has been called the Twelfth-Century Renaissance. Charles Homer Haskins, who pioneered the study of this revival, wrote:

> Unlike the Carolingian Renaissance, the revival of the twelfth century was not the product of a court or a dynasty; and, unlike the Italian Renaissance, it owed its existence to no single country.... The Renaissance of the twelfth century, like its Italian successor three hundred years later, drew its life from two principal sources. Each was based in part upon the knowledge and ideas already present in the Latin West, in part upon an influx of new learning from the East. But whereas the Renaissance of the fifteenth century was concerned primarily with literature, that of the twelfth century was concerned even more with philosophy and science.

The study of natural philosophy in the twelfth-century schools was based principally on Plato's *Timaeus*, the only work of Plato available in western Europe up until the mid-twelfth century, and then only the first fifty-three chapters as translated and commented upon by Chalcidius in the fourth century. As translator and scholar Benjamin Jowett wrote of this curious work: "Of all the writings of Plato, the *Timaeus* is the most obscure and repulsive to the modern reader, and has nevertheless had the greatest influence over the ancient and medieval world."

Plato's ideas in science are contained principally in the *Timaeus*, where he presents a detailed cosmology and cosmogony that Timaeus, his protagonist, says is "along the lines of the likely stories we have been following." Timaeus introduces a divine creator called the *demiourgos*, or craftsman, who uses the ideal Forms as patterns to shape featureless preexisting matter and steer its chaotic motion so as to give order to the cosmos. But, as historian William Guthrie remarked, the *demiourgos* "is not in sole and absolute control, but must bend to his will a material that is to some extent recalcitrant. Otherwise, being wholly good himself, he would have made a perfect world."

According to the *Timaeus*, everything in the universe is composed of the four elements—earth, water, air, and fire—all of which are made up of particles so small that they are invisible. The particles of each element have a definite geometrical shape and are mutually transformable, their main masses arranged in concentric spheres with earth in the center followed by water, air, and fire. The fiery sphere extended from the moon to the fixed stars, containing within it the spheres of the sun, moon, and planets, all of which were made of fire.

Plato relates his cosmogony to the presence of mankind, in a luminous passage where Timaeus points out how our sense of number and time comes from our observation of the heavens: "Our ability to see the periods of day and night, of months and of years, of equinoxes and solstices, has led to the invention of number, and has given us the idea of time and opened the path to inquiry into the nature of the universe. These pursuits have given us philosophy, a gift from the gods to the mortal race whose value neither has been or ever will be surpassed."

Medieval scholars were faced with the task of reconciling Plato's cosmology, as expressed in the *Timaeus*, with the account of creation in Genesis, as explained by the early Church Fathers. This proved to be treacherous territory, as the Church was the only benefactor of science at the time, and acceding to pagans might be viewed harshly by the Church's contemporary leaders. Thierry of Chartres, who flourished in the first half of the twelfth century, toed this line with admirable aplomb.

Thierry was born in Brittany and is believed to have been teaching as early as 1121 at the cathedral school of Chartres, together with his brother Bernard, who was chancellor there in the years 1119–1126. Thierry is recorded as being archdeacon of Dreux, near Chartres, in 1127, and before 1134 he is known to have taught in Paris. He later returned for a time to Chartres, where he is recorded as being "chancellor and archdeacon of Notre-Dame [of Chartres]," to which he bequeathed his *Eptatheuchon*, or *Book of the Seven Liberal Arts*, a summary of the learning of his age in two huge volumes, in which he sought to bring together the *trivium* and *quadrivium* "for the multiplication of the noble tribe of scholars." He seems to have retired around 1155 to a Cistercian monastery, where he died and was buried.

According to scholar Nikolaus Haring, Thierry "is considered to have introduced the concept of *rota* or zero into European mathematics." This was tremendously important, for the addition of zero to the Hindu-Arabic numerals in the form that they had taken in the West led the rapid rise of European mathematics, particularly in number theory, where Latin mathematicians soon began to surpass their Arabic and Greek predecessors.

Thierry's best-known work is the *Hexaemeron*, a short commentary on the introductory chapters of Genesis, in which he tries to give a rational explanation of the six days of creation, based on Plato's *Timaeus* and ideas from the Stoics, Augustine, and Aristotle. According to Thierry, time began with God's creation of the four elements—earth, water, air, and fire—which through their inherent properties and natural law evolved into the material universe, all of this taking place in successive steps during the first six rotations of the heavens. Thierry's cosmology and cosmogony (the creation of the cosmos) are based on his theological interpretation of Aristotle's four causes, which he identifies with the three persons of the Trinity plus matter. God the Father is the efficient cause, the Son is the formal cause, the Holy Spirit is the final cause, and the four elements are the material cause. According to Thierry, the Creator implanted "seminal causes" in the elements to regulate the passage of time, the succession of seasons, and the process of procreation, with a divine power that he calls the world soul governing all matter to give it form and order. He says that this act of orderly creation was due through the Creator's wisdom and solely because of divine benevolence and love.

Several of Thierry's colleagues at Chartres were deeply influenced by his Platonism, most notably William of Conches (c. 1090–after 1154) and Bernard Silvestre (c. 1085–1178), both of whom contributed interesting ideas to the development of medieval European science.

William of Conches was hired around 1145 by Geoffrey Plantagenet to tutor the future king Henry II of England. Before then he had probably begun work on his *De philosophia mundi*, based on Plato's *Timaeus* and other sources. Here William adopted a form of atomism based on a combination of Plato's ideas with those of Lucretius. He stated that God's universe acts according to natural law, saying that the philosopher's task is to understand and explain these laws. He criticized the ignoramuses who fall back on divine intervention and condemned as heretical those who try to give a rational explanation of nature.

Bernard Silvestre is best known for his *Cosmographica*, a poetical cosmogony alternately in prose and verse dramatically describing the six days of creation dedicated to Thierry of Chartres, which he presented to Pope Eugene III in 1147, the only certain date in his life.

Bernard and the other scholars of his time believed in the progress of knowledge, which he and his colleagues at Chartres and elsewhere had achieved through their free and rational use of the learning of their predecessors. As he wrote, in a statement that would be famously echoed by Newton more than five centuries later: "We are like dwarfs standing on the shoulders of giants, so that we can see more things than them, and can see further, not because our vision is sharper and our stature higher, but because we can raise our selves up because of their giant stature."

Another clear indication of the intellectual revival in the early twelfth century is the large number of manuscripts of that period or shortly before concerned with arithmetical and astronomical reckoning. The arithmetical manuscripts are mostly treatises on the abacus, carrying on in the tradition of Gerbert d'Aurillac. Most of the astronomical writings are copies and excerpts from Bede or, occasionally, Isidore of Seville, sometimes in new compilations.

By the second half of the twelfth century the cathedral schools at Chartres and elsewhere had given way to the new universities that were starting to emerge in Europe, brought into being by the great revival of learning and expansion of knowledge, along with the dramatic increase in number of those seeking a higher education. As Haskins wrote in his pioneering work, *The Rise of Universities*, in 1923:

> This new knowledge burst the bounds of the cathedral and monastery schools and created the learned professions; it drew over mountains and across the narrow sea eager youths who, like Chaucer's Oxford Clerk of a later day, "would gladly learn and gladly teach," to form in Paris and Bologna those academic guilds which have given us our first and our best definition of a university, a society of masters and scholars.

The society of master and students referred to by Haskins was known as *universitas societas magistorum discipulorumque*, from which the word "university" stems. This society was organized along the same lines of any of the medieval craft guilds, such as that of the carpenters, which were dominated by the master craftsmen and operated under charters that regulated all of their activities for the mutual benefit of the members. The universities were self-governing corporations that gradually secured varying degrees of freedom from local jurisdiction and taxation, often with the patronage of emperors, kings, popes, and archbishops, which allowed them to establish their own standards and procedures.

The earliest of the universities, the medical school at Salerno, had a different early development than all of the other new institutions of higher learning. Greek medicine had never wholly vanished from the south of Italy, where Latin versions of Galen and other ancient medical writers can be traced as early as the tenth century, when Salerno first established itself as a center of the healing art. The medical school at Salerno was well established by the second half of the eleventh century, as evidenced by the introduction of the Arabic translations of Constantine the African into the curriculum under the title of *Ars medicine* or *Articella*. By the twelfth century Salerno had develop its own medical literature. This was primarily in Latin, but not completely so, for Stephen of Pisa wrote in 1127 that "in Sicily and Salerno, where students of such matters are to be found, there are both Greeks and men familiar with Arabic."

Among the other institutions of higher learning, the earliest was the University of Bologna, founded in 1088, followed in turn by those of Paris (c. 1150), Oxford (1167), Salerno (1173, a refounding of the medical school), Palenzia (c. 1178), Reggio (1188), Vicenza (1204), Cambridge (1209), Salamanca (1218), and Padua (1222), to name only the first ten, with another ten founded in the remaining years of the thirteenth century. Twenty-five more were founded in the fourteenth century, and another thirty-five in the fifteenth, so that by 1500 there were eighty universities in Europe, evidence of the tremendous intellectual revival that had taken place in the West, beginning with the initial acquisition of Graeco-Arabic learning in the twelfth century.

In Paris, for example, the largest and most prestigious university in northern Europe, which at its peak had more than twenty-five hundred students, there were four faculties, including one undergraduate faculty in the liberal arts, by far the largest of the four, and graduate faculties in law, medicine, and theology. A student usually entered at age fourteen, having previously learned Latin at a grammar school, enrolling under a master whose lectures he attended for three or four years before taking an examination for the degree of bachelor of arts. This degree permitted him to give certain types of lectures under the direction of a master while continuing his studies. When he had followed lectures in all of the required subjects, he could take the examination for the master of arts degree, and if he passed he was given full membership in the arts faculty, with the right to teach any course in the arts curriculum. A student could then enroll in one of the graduate faculties, often while still teaching in the undergraduate faculty. The graduate programs were long and demanding, taking anywhere from five to fifteen years of additional study before one could take an examination for a doctor's degree in law, medicine, or theology.

The curriculum evolved, so far as the trivium was concerned, with a much greater emphasis on logic at the expense of grammar, while in the quadrivium astronomy dominated, particularly as regards timekeeping and calendrical problems as well as its application to astrology. The liberal arts curriculum was expanded to include lectures in moral philosophy, mental philosophy, and metaphysics, with law, medicine, and theology taught as advanced subjects in the graduate program. The courses were based on textbooks rather than subjects. The teaching texts were Greek and Arabic works in Latin translations, along with new works written especially for the courses. By the second half of the thirteenth century Aristotle's influence had increased to the point that his works on metaphysics, cosmology, physics, meteorology, psychology, and natural history were required reading, along with commentaries, so that all graduates were thoroughly grounded in Aristotelian natural philosophy.

Bologna became the archetype for later universities in southern Europe, Paris, and Oxford for those in the northern part of the continent. Bologna was renowned for the study of law and medicine, Paris for logic and theology, and Oxford for philosophy and natural science. Training in medicine was based primarily on the teachings of Hippocrates and Galen, astronomy and astrology were based heavily on the writings of Ptolemy, and studies in logic, philosophy, and science were based heavily on the works of Aristotle and commentaries upon them, at first translated from Arabic and then later from Greek.

Meanwhile European scholars were absorbing the Graeco-Arabic learning that they had acquired and adapting it to develop a new philosophy of nature, which, although primarily based upon Aristotelianism, differed from some of Aristotle's doctrines right from the beginning.

This is evident in the works of Peter of Abano, who, as we learned, tried to reconcile the different theories, often conflicting, of his predecessors, most notably Aristotle, his most revolutionary suggestion being that the stars were not fixed but moving freely in space. This idea, which would be revived in the seventeenth century, would be the first step in breaking the bounds of the static, finite, and earth-centered Aristotelian cosmos, and paving the way for the dynamic and expanding universe of modern astrophysics that has opened up in my own lifetime.

It is interesting that Peter said that he and his colleagues at the University of Paris had been debating these ideas "for the past decade," clear evidence that the gathering of large numbers of scholars with diverse backgrounds at the new universities was producing an exchange of ideas reminiscent of Plato's Academy, Aristotle's Lyceum, and the Museum and Library of Alexandria.

It is also significant that Peter had in his search for ancient manuscripts gone to Constantinople after the renewal of contacts between Latin West and Greek West, and that there he had met Marco Polo, who told him about the journey that had taken him all the way to China and back. At about the same time Fibonacci was in Constantinople, where, as we have learned, he met with Latin, Greek, and Arabic mathematicians, one of the contacts that led to his transmission of the Hindu-Arabic number system to western Europe.

The horizons of Latin Europe were expanding, both intellectually and geographically, setting the stage for the emergence of modern Europe and of modern science, which would eventually spread through the entire world.

5

Converting Aristotle

*T*HE CURRICULA IN THE NEW EUROPEAN UNIVERSITIES WAS, AS WE have learned, largely devoted to the works of Aristotle, whose ideas in many cases ran counter to Roman Catholic dogma, which led to attempts at reconciling the different views, beginning with that of Peter of Abano.

The problem of accommodating Aristotle's thought to Roman Catholic dogma preoccupied philosophical discussion in western Europe during the thirteenth century. Two scholars in particular dominated this discussion: Albertus Magnus and his even more famous student Thomas Aquinas. As one of their contemporaries wrote of them: "Doubtless many others were famous during this same time both in life and thought. But these two transcended and deserve to be placed above all others."

Albertus Magnus (c. 1200–1280) was born to a family of the military nobility in Bavaria. He studied liberal arts at the University of Padua, where he was recruited into the Dominican order by its master general, Jordanus of Saxony. He then studied theology and taught in Germany before enrolling in the University of Paris in around 1241, where he lectured on theology for seven years before he was sent to open a school in Cologne. His students included Thomas Aquinas, who came from Italy to study with him, either in Paris or Cologne. Albertus was appointed provincial of the German Dominicans in 1253 and in 1260 he became bishop of Regensburg, a post that he reigned two years later, after which he spent the rest of his life preaching and teaching. He took part of the Second Council of Lyons in 1274, and three years later he went to Paris. There he tried to stop the condemnation of Aristotle's doctrines by Pope John XXI, in which some of the ideas of his student Thomas Aquinas were questioned, particularly concerning the absolute power of God, which theologians felt was being limited by Thomas.

Albertus seems to have begun studying the works of Aristotle when he first came to teach in Paris. It was probably then that he began his monumental compendium of all the works of Aristotle known at the time, as well as those of the so-called Pseudo-Aristotle. Together these make up seventeen of the forty volumes in the critical edition of Albertus's works. He undertook this task at the request of his Dominican brethren, who wanted to explain, in Latin, the principal Aristotelian physical doctrines. As he wrote in the prologue to his commentary on Aristotle's *Physics*, his purpose was "to make all parts of philosophy intelligent to the Latins."

He went far beyond the request of his brethren, explaining not only the natural sciences but also mathematics, logic, metaphysics, ethics, and politics, adding in his commentaries all that he had learned of Graeco-Arabic science and philosophy. His synthesis also included what he knew of Plato's thought as well as his comments on a number of Neoplatonic writings. His acceptance of the basic Aristotelian system is clear in his rejection of Platonic and Pythagorean ideas in cosmology and natural philosophy. Nevertheless, he does not hesitate to disagree with Aristotle or Aristotelian commentaries when he thinks they are wrong. As he wrote, disagreeing with the view of some contemporaries that Aristotle was infallible: "Whoever believes that Aristotle was a god, must also believe that he never erred. But if one believes that he was a man, then doubtless he was liable to error just as we are." Albertus's *Summa theologica*, for instance, contains a listing of Aristotle's errors, and in his *Meteorology* he notes at one point that "Aristotle must have spoken from the opinions of his predecessors and not from the truth or demonstration of experiment."

He believed that sound was caused by the impact of two hard bodies, producing vibrations that are propagated spherically. He performed simple experiments to see how sunlight produced thermal effects, such as showing that a black object will become hotter than a mirror, and he speculated, correctly, that refraction of rays from the sun played a role in the formation of rainbows. Writing on this matter, he corrected Aristotle's statement that a lunar rainbow occurs only twice in fifty years, noting, "I myself have observed two in a single year."

Through his researches and writing, Albertus played a crucial role in rediscovering Aristotle and making his philosophy of nature acceptable to the Christian West. The main problem involved in the Christian acceptance of Aristotle was the conflict between faith and reason, particularly in the Averroist interpretation of Aristotelian philosophy with its determinism and its view of the eternity of the cosmos. Albertus sought to resolve this conflict by regarding Aristotle as a guide to reason rather than an absolute authority, and where he conflicted with either revealed religion or observation, then he must be wrong. He held that natural philosophy and theology often spoke of the same thing in different ways, and so he assigned to each of them its own realm and methodology, sure that there could be no essential contradiction between reason and revelation.

One of the objections to Aristotle by theologians was that his system limited the power of God, whom they believed was omnipotent and could thus, for example, have created a multiplicity of universes. Albertus, following Aristotle, concluded that "it is impossible that there be several worlds," adding that he was referring only to what is impossible in nature, for "there is a great difference between what God can do by means of his absolute power and what can be done in nature." He felt that natural science should not be concerned with what God is capable of doing, but only with what he actually has done, for after the act of creation the world functions "according to the inherent causes of nature."

Both Albertus and Aquinas differed from their two famous contemporaries at Oxford, Robert Grosseteste and Roger Bacon, in that they were more purely Aristotelian in their philosophy, whereas the latter took a more Platonist view. Grosseteste and Bacon believed that the principles of natural science are essentially mathematical, while Albertus held that mathematics is an abstract science whose application must be evaluated by the science that studies nature as it actually exists, "in motion and in concrete detail."

Albertus's ideas concerning motion and gravitation are essentially Aristotelian. He mentions the term *impetus* when discussing projectile motion, but refers to it as coming from the medium rather than residing in the moving body, following Aristotle. He also follows Aristotelian theory in explaining the acceleration of a falling body, saying that it speeded up as it approached earth because of its increasing desire to be in its natural place.

Albertus also speculated that the Milky Way is made up of a multiplicity of stars and that the dark areas on the moon are due to surface configurations and not the earth's shadow, ideas that would be verified by Galileo in his first observations with a telescope. Albertus's treatise on comets makes use of simple observations to verify or reject theories that had been proposed to explain them, correctly concluding that they were celestial objects moving in the earth's atmosphere. He followed earlier thinkers in explaining tides as being due to the motion of the moon and favored the Ptolemaic theory of planetary motion. He was aware of the precession of the equinoxes, though he wrongly attributes its discovery to Aristotle rather than Hipparchus.

The most original contributions made by Albertus are in botany and the life sciences, where his work was distinguished by his acute observations and skill in classification. He was the first Latin scholar known to have made use of what became the modern scientific method, the combination of theory, in his case Aristotelian syllogisms, and experimental observation. This is evident in his *De vegetabilibus et plantis*, an encyclopedic commentary on the pseudo-Aristotelian *De plantis*, which was the principal source of botanical knowledge in Latin Europe down to the sixteenth century. Discussing the native plants known to him, he wrote: "In this sixth book we will satisfy the curiosity of the students rather than philosophy.... Syllogisms cannot be made about particular natures, of which experience (*experimentum*) alone gives certainty," where he was referring to his observations of particular plants as compared to conclusions regarding them arrived at through Aristotle's teleological theory.

Writing of Albertus's "digressions" in *De vegetabilibus*, science historian A. C. Crombie remarked that they "show a sense of morphology and ecology unsurpassed from Aristotle and Theophrastus to Cesalpino and Jung."

Albertus followed the main outlines of the classificatory scheme that Theophrastus had laid out in his *Inquiry into Plants*, in which plants were classified into trees, shrubs, undershrubs, and herbs, with finer distinctions such as those between cultivated and wild, flowering and flowerless, fruit-bearing and fruitless, deciduous and evergreen.

The appearance of new species had been the subject of speculation since the time of the first Ionian philosophers of nature. Anaximander thought that all life had originated by spontaneous generation from water and that mankind had developed from fish. Most other ancient writers held that the succession of new species was generated from a common source such as the earth, rather than by modification by living ancestors. Albertus followed Theophrastus in believing that existing types were sometimes mutable, describing five ways in which one plant can be transformed to another, including grafting, the domestication of wild plants, and the running wild of cultivated plants.

Albertus's *De animalibus* is a good example of the way in which he and other medieval natural philosophers used the translations and commentaries on the works of Aristotle and other Greek writers to make their own observations and give modified explanations. The first nineteen of the twenty-six books of *De animalibus* are a commentary on Aristotle's *History of Animals*, *Parts of Animals*, and *Generation of Animals*, all in Michael Scot's translation. Albertus's commentary also makes use of ibn Sina's own commentaries on these works, of ibn Sinas's *Canon*, based on Galen, and of Latin translations of some of Galen's works. The last seven books of *De animalibus* consist of original discussions by Albertus on various biological topics, as well as descriptions of particular animals, taken partly from Thomas of Cantimpré's *De rerum natura* (c. 1228–1244).

Thomas of Cantimpré's *De rerum natura* contains a description of herring fisheries and the hunting of seals, walruses, and whales, as well as a section on fabulous animals. This was one of a number of encyclopedias on agriculture and nature that appeared in the thirteenth century and the first half of the following century, others being treatises on animal husbandry by Walter of Henley and Peter Crescenti, the sections on agriculture in the encyclopedias of Albertus's *De vegetalibus* and the *Speculum Doctrinale* of Vincent of Beauvais, and the botanical and zoological sections in Bartholomew the Englishman's *The Nature of Things*, which may be the source of Shakespeare's natural history. These encyclopedias are evidence of a lively interest in nature also shown in the art, architecture, and books of the period, as in illuminated and illustrated manuscripts, herbals, paintings, mosaics, and reliefs, as well as in the menageries and zoological gardens kept by kings, princes, and even towns. These were not just hobbies and elite studies, but more the result of a renewed interest in nature in western Europe during the late medieval era, leading to the revival of the life sciences in the seventeenth century.

Albertus's own researches in embryology are described in *De animalibus* as digressions in his commentary on Aristotle's *Generation of Animals*. There, in book 5, he gives a remarkable description of the life history of a butterfly or moth based on his own observation.

He also gave excellent descriptions of a large number of northern animals unknown to Aristotle, noting, for example, the varieties of color of the squirrel, changing from red in Germany to gray in Russia, and the lightening in hue of falcons, jackdaws, and ravens in cold climates.

Albertus's geology comes mostly from Aristotle's *Meteorologica*, the Pseudo-Aristotelian *De elementiis*, and Avicenna's *De mineralibus*, but he adapted these authorities to formulate a coherent theory and added observations of his own. He extended Avicenna's account of fossils, for example, giving his own explanation in his *De mineralibus et rebus metallica*: "There is no-one who is not astonished to find stones which, both externally and internally, bear the impression of animals. Externally they show their outline and when they are broken open there is found the internal parts of these animals. Avicenna teaches us that the cause of this phenomenon is that animals can be completely transformed into stones and particularly into salt stones." This did not create any problems with the Church, for there was nothing in this work that would seem to be in conflict with its dogmas.

Albertus was also interested in alchemy and performed a number of chemical experiments, compiling a list of the properties of some one hundred minerals. His theory of the structure of matter includes the concept of elements in compounds, and he is said to have been the first to isolate the element arsenic.

Although Albertus was very modern in his scientific thinking, he was still medieval in his views on such matters as magic, divination, and astrology. He wrote in his *Summa theologica* of his belief that magic is due to demons. "For the saints expressly say so, and it is the common opinion of all persons, and it is taught in that part of necromancy which deals with images and rings and mirrors of Venus and seals of demons." Albert writes of astrology in almost all of his scientific treatises, describing the effects produced by such celestial phenomena as conjunctions of the planets, to which he attributes "great accidents and great prodigies and a general change of the state of the elements and of the world."

Ulrich Engelbert of Strasburg, a pupil of Albert's, described him as "a man in every science so divine that he may well be called the wonder and miracle of our time." Thomas Aquinas wrote of him with equal admiration, saying, "What wonder that a man of such whole-hearted devotion and piety should show superhuman attainments in science." Albert was canonized by Pope Pius XI on December 16, 1931, and ten years later Pope Pius XII declared him the patron saint of all those who cultivate the natural sciences.

Thomas Aquinas (c. 1225–1274) was born near Monte Cassino in southern Italy, where his father served the emperor Frederick II in his war against the papacy. He began his education at the Benedictine abbey of Monte Cassino, after which he went to the newly founded University of Naples, where he was introduced to the works of Aristotle. After joining the Dominicans, he was sent for further studies to Cologne and then Paris, where his teachers included Albertus Magnus. Studying under Albertus, Thomas soon mastered the most recent scholarship of his time, including the major Greek and Arabic works that had been translated into Latin.

Thomas spent two periods as professor at the University of Paris, 1256–1259 and 1269–1272, and in the interim he was associated in turn with the papal courts of Alexander IV, Urban IV, and Clement IV. After his second professorship in Paris he returned to Naples to start a Dominican school, which he directed until a few months before his death in 1273. He was canonized by Pope John XXII on July 18, 1323, and was subsequently presented by the Roman Catholic Church as the most representative teacher of its doctrines. His philosophical system, known as Thomism, is still taught at Catholic universities, and I first studied it as an undergraduate at Iona College in New Rochelle, New York.

Thomas continued Albertus's program of assimilating Aristotle's philosophy and adapting it to Roman Catholic dogma as he interpreted it. His own synthesis of pagan and Christian thought is definitively expressed in his massive *Summa theologica*, which he was still working on at the time of his death. One of the questions he addresses in this work is the difference between a hypothesis that must necessarily be true, such as a physical or metaphysical hypothesis, and one that merely fits the observed facts, such as a mathematical hypothesis. An example of a metaphysical hypothesis was the Aristotelian view that the celestial bodies are embedded in a set of concentric crystalline spheres rotating around the earth with constant velocity. A mathematical hypothesis, on the other hand, was the Ptolemaic system of eccentrics and epicycles to fit the observed motions of the sun, moon, and planets.

As it turns out, both of these hypotheses are false, as Copernicus, Galileo, Kepler, and Newton were to show. But Thomas thought that a physical or metaphysical hypothesis, such as Aristotle's view that celestial bodies rotate around the earth with constant velocity, must necessarily be true because "sufficient reason can be brought to show that the motions of the heavens are always of uniform velocity." He went on to say that a mathematical hypothesis, such as the Ptolemaic theory of cycle and epicycles, "is not a sufficient proof, because possibly another hypothesis might be also be able to account for them."

Thomas, like Albertus Magnus, tried to resolve the conflict between theology and natural science and show that there could be no real contradiction between revelation and reason. Arguing against those who said that natural philosophy was contrary to the Christian faith, he wrote in his treatise on *Faith, Reason, and Theology* that "even though the natural light of the human mind is inadequate to make known what is revealed by faith, nevertheless what is divinely taught to us by faith cannot be contrary to what we are endowed with by nature. One or the other would have to be false, and since we have both of them from God, he would be the cause of our error, which is impossible."

Many of the problems involved in adapting Aristotle to Catholic dogma were addressed by Thomas in two of his books: *On the Eternity of the World* and *On the Unicity of the Intellect Against the Averroists*. Thomas, in the first of these books, says that we know from divine revelation that the world was created at a moment in time, but philosophy cannot settle the problem one way or the other, since the question of creation ex nihilo is one that cannot be proved by using natural reasons but depends on faith alone. The second book presents Thomas's arguments against the Averroist concept revived by Siger of Brabant, a radical lecturer at the University of Paris, of monopsychism, that is, the notion that the human soul is not confined to a single person but is a unitary intellect shared by all humans, and what survives after the death of the individual is not personal but collective.

After Thomas published his work *On the Unicity of the Intellect*, Siger modified his views to conform them to Catholic teaching. Nevertheless, Siger insisted that his ideas were necessary philosophical conclusions, but since they conflicted with Catholic views, the dogma of the Church must prevail. Siger wrote: "One should not try to investigate those things which are above reason or to refute arguments for the contrary position. But since a philosopher, however great he may be, may err on many points, one ought not to deny the Catholic faith because of some philosophical argument, even though he does not know how to refute it."

Boethius of Dacia (fl. 1270), a member of Siger's circle, wrote the treatise *On the Eternity of the World*, in which he set out to refute the notion of creation, arguing in favor of the Aristotelian idea of a universe that had no beginning in time. Having done so, he made it clear that he himself, as a Christian, accepted the doctrine of creation as a matter of faith. Nevertheless, he argued that the philosopher is bound to examine any matter that

lends itself to rational explanation. As he wrote: "It belongs to the philosopher to determine every question which can be disputed by reason; for
every question which can be disputed by rational argument falls within
some part of being. But the philosopher investigates all being—natural,
mathematical, and divine. Therefore it belongs to the philosopher to determine every question which can be disputed by rational argument."

Although Aristotle's works formed the basis for most nonmedical studies at the new universities, some of his ideas in natural philosophy, particularly as interpreted in commentaries by Averroës, were strongly opposed
by Catholic theologians. One point of objection to Aristotle was his notion
that the universe was eternal, which denied the act of God's creation; another was the determinism of his doctrine of cause and effect, which left no
room for divine intervention or other miracles. Still another objection was
that Aristotle's natural philosophy was pantheistic, identifying God with
nature, which derived from the Neoplatonic interpretation of Aristotelianism by Avicenna (ibn Sina).

This led to a decree, issued by a council of bishops at Paris in 1210,
forbidding the teaching of Aristotle's natural philosophy in the university's
faculty of arts. The ban was renewed in 1231 by Pope Gregory IX, who issued a bull, or papal decree, declaring that Aristotle's works on natural philosophy were not to be read at the University of Paris "until they shall have
been examined and purged from all heresy." The ban was apparently not
enforced, and in any event it seems to have remained in effect for less than
half a century, for a list of texts used at the University of Paris in 1255 includes all of Aristotle's available works.

The controversy was renewed in 1270, when the bishop of Paris, Eti-
enne Tempier, condemned thirteen propositions derived from the philosophy of Aristotle or from Aristotelian commentaries by Averroës. This gave
rise to the notion of "double truth," in which an idea might be true if
demonstrated by reason in physics and metaphysics, while a contradictory
concept could be independently true in theology and the realm of faith.
Pope John XXI, after seeking the advice of theologians, issued a bull in
1277 in which he condemned 219 propositions, including the original 13

listed by Tempier, threatening excommunication of anyone who held even a single erroneous doctrine. That same year, a similar condemnation was issued by Tempier as well as by the archbishop of Canterbury, Robert Kilwardby, whose edict was renewed in 1284 by his successor, John Pecham. A number of the propositions were declared to be erroneous because their determinism placed limits on the power of God.

The condemnation of Averroist doctrines by the bishop of Paris in 1270 may have been directed against some of the teachings of Thomas Aquinas, one being that the creation of the world cannot be demonstrated by reason alone. This and other interpretations by Thomas were his solution to the problem of adapting Aristotelianism to Christian theology. The lengths to which Thomas went is evident in his attempt to fit the biblical account of the Ascension into the Aristotelian cosmos. According to Ephesians 4:10, Christ "ascended up far beyond all heavens, that he might fill all things," which presented problems for Thomas in trying to square this with Aristotle's philosophy and his model of the homocentric crystalline spheres.

Such were the efforts that Thomas made to adapt Aristotle's system of the world to Catholic theology. Despite the condemnations that were made in the thirteenth century, the works of Aristotle continued to dominate the curriculum at Paris and other universities. A regulation adopted at the University of Paris in 1341 required all new masters of arts to swear that they would teach "the system of Aristotle and his commentator Averroës and of the other ancient commentators and expositors of the said Aristotle, except in those cases that are contrary to the faith." By that time Aristotelian philosophy had become the basis for undergraduate studies and in the graduate programs in medicine, law, and theology in all European universities, as well as providing the foundation for philosophical discussion and scientific research. This was the culmination of what can be termed the conversion of Aristotle to Christianity, which historian David Lindberg credits largely to the effort of Thomas Aquinas in solving the problem of faith and reason.

This, then, is Thomas's solution to the problem of faith and reason.

Above: A drawing of Saint Albertus Magnus by Tommaso da Modena

Right: A painting of Saint Thomas Aquinas by Fra Angelico

He had made room for both, subtly merging Christian theology and Aristotelian philosophy into what we may call "Christian Aristotelianism." In the process it was necessary for Thomas to Christianize Aristotle by wrestling with the Aristotelian doctrines that appeared to conflict with the teaching of revelation and correcting Aristotle where he had fallen into error; at the same time he "Aristotelianized" Christianity, importing major portions of Aristotelian metaphysics and natural philosophy into Christian theology. In the long run Thomism came to represent the official position of the Catholic Church; in the short run Aquinas was viewed by theologians of more conservative persuasion as a dangerous radical.

Albertus Magnus and Thomas Aquinas had in effect converted Aristotle to Christianity, so that Aristotelianism, with its static and earth-centered cosmology, represented the worldview of western Europe up until the seventeenth century, by which time the works of Copernicus, Galileo, Kepler, and Newton brought about its downfall.

Meanwhile Aristotelianism remained the basis of higher education and scientific research in western Europe, where at the beginning of the thirteenth century scholars would begin to develop a new philosophy of nature and a scientific method based on observation and experiment, while at the same time their thinking was still rooted in the works of Aristotle, as they walked a tightrope to avoid conflict with Church dogma.

6

The Metaphysics of Light

*T*HE NEW EUROPEAN PHILOSOPHY OF NATURE WOULD EMERGE IN the universities of Paris and Oxford, where the intellectual ferment of the diverse student body produced revolutionary ideas that, as we have seen, occasionally brought down on them the heavy hand of the Church, though never for long. The intellectual revival of western Europe was now unstoppable, though even the rebels were still rooted in the ideas of Aristotle, revised so as not to run counter to Catholic dogma, as in the Thomistic philosophy I studied as an undergraduate.

The new philosophy of nature would be based on theory and experiment, as well as mathematics, though at this stage scientists in Europe had basically the same equipment as did the Greeks of the Hellenistic period, when Ptolemy used an astrolabe for his astronomical observations and simple optical instruments for his study of light, so that their experimentation was restricted to optics. Nevertheless, their studies of light, which they extended to acoustics, the science of sound, produced results that in some cases went beyond the level achieved by their Greek and Arab predecessors, setting the stage for the emergence of modern optics in the seventeenth century. They also went beyond their predecessors in the study of the refraction of light by lenses, leading to the invention of the telescope and the microscope.

One of the most influential figures in the rise of the new European philosophy of nature was Robert Grosseteste (c. 1175–1253). His biographer, A. C. Crombie, calls him "the real founder of the tradition of scientific thought in medieval Oxford, and in some ways, of the modern English intellectual tradition."

Born of humble parentage in Stradbroke in Suffolk, England, he was educated at the cathedral school in Lincoln and then at the University of Oxford. Grosseteste was in the household of William de Vere, bishop of Hereford, by 1198, when a reference by Gerald of Wales suggests that he may have had some competence in both law and medicine, with a manifold learning "built upon the sure foundation of the liberal arts and an abundant knowledge of literature."

This comment is substantiated by what is probably Grosseteste's first work, *De artibus liberalibus*. In the introduction he describes how the seven liberal arts acted as a purgative of errors and gave direction to the mind. His treatment of music is particularly interesting, since for him the laws of harmony applied not only to the human voice and musical instruments but also to the movement of the celestial bodies, the composition of bodies made up of the four terrestrial elements, and the harmonic relation between body and soul in man. He also wrote a related essay entitled *De generatione sonorum*, in which he describes sound as a vibratory motion propagated from the sounding body through the diaphragm of the ear, whose motion arouses a sensation in the soul.

After that Grosseteste probably taught in the arts faculty at Oxford until 1209, when the masters and scholars of the university were dispersed for five years because of a particularly violent student riot in the town. During those years he received a master's degree in theology, probably at the University of Paris. At some time in the period 1209–1214 he was appointed *magister scholarum*, or chancellor of the University of Oxford, probably the first, or one of the first, to hold this office. He also would have lectured on theology, while apparently beginning his own study of Greek. When the first Franciscan monks came to Oxford in 1224, Grosseteste was appointed as their reader, and he directed their interests toward mathematics and natural science as well as the study of the Bible and languages. He finally left the university in 1235 when he was appointed bishop of Lincoln, his jurisdiction including Oxford and its schools. During his episcopate he attended the First Council of Lyons in 1245. He died on December 9, 1253, in Buckden in Buckinghamshire, and he was buried in the cathedral at Lincoln.

Grosseteste's writings are divided into two periods, his chancellorship of Oxford and his tenure as bishop of Lincoln. His writings in the first period include his commentaries on Aristotle and the Bible and most of his independent treatises. Those in the second period are principally his translations from the Greek: Aristotle's *Nicomachean Ethics* and *On the Heavens*, the latter along with his version of the commentary by Simplicius, as well as several theological works. He brought to Lincoln scholars who knew Greek to assist him in his translations. He also arranged for a translation of the Psalms to be made from the Hebrew, and he seems to have learned something of that language.

The commentaries that Grosseteste wrote on Aristotle's *Posterior Analytics* and *Physics* were among the first and most influential interpretations of those works. These two commentaries also presented his theory of science and scientific method, which he put into practice in his own writings, including six works on astronomy and one on calendar reform, as well as treatises entitled *The Generation of the Stars, Sound, The Impressions of the Elements, Comets, The Heat of the Sun, Color, The Rainbow*, and *The Tides*, in which he attributed tidal action to the moon.

Grosseteste was the first medieval European scholar to use Aristotle's methodology of science. Grosseteste's methodology involved two steps. The first of these was a combination of induction and deduction, which he called resolution and composition, respectively. The second step was what Grosseteste called verification and falsification, a process necessary to distinguish the true cause from other possible causes. He based his use of verification and falsification upon two assumptions about the nature of physical reality. The first of these was the principle of the uniformity of nature, in support of which he quoted Aristotle's statement that "the same cause, provided that it remains in the same condition, cannot produce anything but the same effect." The second was the principle of economy, which holds that the best explanation is the simplest, that is, the one with the fewest assumptions, other circumstances being equal. Here again he quoted Aristotle, who said that power from natural agents proceeds in a straight line "because nature operates in the shortest way possible." Beginning with these assumptions, Grosseteste's method was to distinguish between possible causes "by experience and reason," rejecting theories that contradicted either factual evidence or an established theory verified by experience. These ideas led to the scientific method that was used by Galileo and Newton and others to establish the foundations of modern science.

The experiments performed by Grosseteste were principally in optics, studying the reflection of light by plane and curved mirrors and its refraction by lenses as well as a spherical glass container full of water. His experiments on refraction gave him an understanding of color, anticipating some of the discoveries made by Newton and published early in the eighteenth century in his *Optiks*.

The methodology of Grosseteste also used Aristotle's procedure of subordinating some sciences to others, such as of astronomy and optics to geometry and of music to arithmetic. As he wrote in one of his Aristotelian commentaries: "With such sciences of which one is under the other, the superior science provides the *propter quid* [the reason the fact] for that thing of which the inferior science provides the *quia* [the observed fact]."

According to Grosseteste, it was impossible to understand the physical world without mathematics, which, in my opinion, sets him apart as the first modern physicist. Although mathematics involved a study of abstract quantity, mathematical entities actually existed as quantitative aspects of physical things, for, as he noted, "quantitative disposition are common to all mathematical sciences ... and to natural science." The use of mathematics made it essential to perform measurements that have a quantitative result, though in doing so there was an inescapable inaccuracy, which made all human measurements conventional. But although geometry, for example, could give the "reason for the fact," in the sense of describing a phenomenon in optics such as reflection of light, it could not provide the efficient and other causes involved. Thus a complete explanation of optical phenomena requires not only geometry, but a knowledge of the physical nature of light that causes it to move as it does in being reflected by a mirror, in which the angle of incidence equals the angle of reflection. That is, mathematics provided only the formal cause; the material and efficient causes were provided by the physical sciences. Thus "the cause of the equality of the two angles made on a mirror by the incident ray and the reflected ray is not a middle term taken from geometry, but is the nature of the geometry generating itself in a straight path."

Although many of Grosseteste's ideas were Aristotelian in origin, some of them differed significantly from those of Aristotle. Aristotle, for example, held that all of the celestial bodies were composed of the quintessential element, aether, while Grosseteste believed that the stars were made up of the four terrestrial elements. Also, whereas Aristotle argued that a vacuum was impossible and that space is finite in extent, Grosseteste tried to explain the meaning that could be given to the mathematical concepts of a vacuum and infinite space. He suggested that space "as it was imagined by mathematicians" could be thought of as infinite in extent only because it was not the same as real space.

He explains how mathematics and measurement provided a means of describing natural phenomena. Commenting on Aristotle's definition of time as "the number of movement in respect of before and after," he suggested that by using this definition, rates of local motion and other kinds of change could be compared by measuring lengths.

He goes on to say that mathematical physicists must, for practical reasons, measure magnitudes of all kinds by conventional units, such as those of the finger, span, and cubit for length, and "one revolution of the heavens" as a measure of time. Here he is reviving the Platonic notion of "geometricizing nature," now extended to quantifying even subjective aspects of nature as well as those that are not eternal but change in time. As Crombie has noted, this "foreshadowed in a striking manner a methodological principle on which modern mathematical physics, particularly since the seventeenth century has been based ... the principal that, in order to be described in the language of science, 'subjective' sensations should be replaced by concepts amenable to mathematical treatment."

It was Grosseteste's belief that the study of optics was the key to an understanding of the physical world, a notion that stemmed from his Neoplatonic "Metaphysics of Light." Following the lead of Saint Augustine, he held that physical light is analogous to the spiritual light by which the mind received true knowledge of the ideal forms that he thought to be the essential principals in the order of nature.

He felt that something was known with complete certitude when the concept of it corresponded to the eternal form existing in God's mind. In one of his Aristotelian commentaries Grosseteste says of Aristotle's statement that "the science is more certain and prior which is knowledge at once of the fact and of the reason for the fact."

According to Grosseteste, in the beginning God created light, the fundamental corporeal form, which multiplied itself infinitely in every direction and in its expansion forced unextended matter into spatial dimensions. He considered light to be the efficient cause of motion and the coordinating principle that gives order and intelligibility to the macrocosm of the created universe as well as governing the interaction between soul and body and the bodily senses in the microcosm of man.

According to his optical theory, light travels in a straight line through the propagation of a series of waves or pulses, and because of its rectilinear motion it can be described geometrically. This was similar to the acoustical theory he presented in his commentary on Aristotle's *Posterior Analytics*, where he wrote that "when the sounding body is struck and vibrating, a similar vibration and similar motion must take place in the surrounding contiguous air, and this generation progresses in every direction in straight lines."

He thought that the same theory, which he called the multiplication of species, could be used to explain the propagation of any disturbance, be it light, sound, heat, mechanical action, or even astrological influence. Thus the study of light was of crucial importance for an understanding of nature. He also believed that light, by which he meant not only visible radiation but the divine emanation as well, was the means by which God created the universe.

The Metaphysics of Light

The study of optics was divided by Grosseteste into three parts, namely phenomena involving vision, mirrors (catoptrics, or reflection), and lenses (dioptrics, or refraction). He discussed the third part more fully than the other two, noting that it had been "untouched and unknown among us until the present time," and suggested applications of refraction that in the seventeenth century would be realized through the invention of the telescope and the microscope. "This part of optics," he wrote, "when well understood, shows us how we may make things a very long distance off appear as if placed very close… and how we may make small things placed at a distance appear any size we want, so that it may be possible for us to read the smallest letters at incredible distances, or to count sand, or grains, or seeds, or any sort of minute objects."

The reason for this magnification, as he had learned from the optical treatises of Euclid and Ptolemy, was "that the size, position and arrangement according to which a thing is seen depends on the size of the angle through which it is seen and the position and arrangement of the rays, and that a thing is made invisible not by great distance, except by accident, but by the smallness of the angle of vision." Thus "it is perfectly clear from geometrical reasons how, by means of a transparent medium of known size and shape placed at a known distance from the eye, a thing of known distance and known size will appear according to place, size and position."

Grosseteste developed a quantitative theory of refraction in an attempt to explain the focusing of light by a "burning-glass," or spherical lens. According to his law, which is incorrect, when light passes from a dense medium to a rarer one, the refracted ray bisects the angle between the incident ray and the perpendicular to the common surface at the point of entry.

An experiment would have shown Grosseteste that his law of refraction was incorrect, but apparently he never put his law to the test, although it was one of the basic tenets of his scientific method that if a theory was contradicted by observation, it must be abandoned. Crombie explained this curious attitude:

He was, in fact, primarily a methodologist rather than an experimentalist, and also, perhaps, he was too much obsessed with the principle, according to which he believed *lux* [the essence of visible light] to behave, and with the alleged similarity between refraction and reflection, to arrive at a correct understanding of the problem.

Nevertheless, Grosseteste's proper application of his scientific method is evident in his treatise on *The Rainbow*, in which he broke with Aristotelian theory by holding that the phenomenon was due to refracted rather than reflected light. As he writes there, using his notion of the subordination of some sciences to others: "For example, optics falls under geometry, and under optics falls the science concerned with the rays of the sun refracted in a concave watery cloud." Although his theory of the rainbow was incorrect, he posed the problem in such a way that investigations by those who followed after him approached closer to the true solution through criticizing his efforts.

He held that it was possible to explain qualitative differences in physical powers as stemming from quantitative differences based on geometrical properties of light rays. Thus he tried to explain the intensity of heat and light as due to the concentration of rays, and heat itself as a scattering of molecular parts due to movement caused by radiation. He said the color was "light incorporated with the transparent medium" and that the entire spectrum of colors was produced by the "intension and remission" of three factors, namely the purity of the medium from earthy matter, the brightness of the light, and the quantity of the rays. He concluded: "That the essence of color and a multitude of the same behaves in the said way is manifest not only by reason but also by experiment, to those who know the principle of natural science and of optics deeply and inwardly.... They can show every color they wish to visibly, by art [*per artificium*]."

There is little in Grosseteste's philosophical and scientific writings to indicate that he was a Christian bishop, but in his treatise *On the Fixity of Motion and Time* he differed from the Aristotelian doctrine that the universe is eternal, for that contradicted his belief in God's creation. His Christian beliefs are also evident in another treatise, *On the Order of the Emana-tion of Things Caused from God*, in which he says that he wishes men would cease questioning the biblical account of the creation.

Beside his works on optics, Grosseteste wrote a number of treatises on astronomy. The most important of these was *De sphaera*, in which he discussed elements of both Aristotelian and Ptolemaic theoretical astron-omy. He also wrote of Aristotelian and Ptolemaic astronomy in four trea-tises on calendar reform, where he used Ptolemy's system of eccentrics and epicycles to compute the paths of the planets, though he noted that "these modes of celestial motion are possible, according to Aristotle, only in the imagination, and are impossible in nature, because according to him all nine spheres are concentric."

His works on calendar reform stemmed from his observation that the system that had long been in use—a cycle in which nineteen solar years were equal to 235 lunar months—was in error, as evident from the fact that the moon was never full when the calendar said it should be, which meant that the reckoning of Easter was always wrong. This was due to the inac-curacy both in taking the solar year to be 365.25 days and in having the lunar cycle equal to nineteen years.

To correct the problem, Grosseteste's plan for calendar reform in-volved three stages, the first of which was an accurate determination of the solar year. He was aware of three such measurements, that of Hipparchus and Ptolemy, and those of Thabit ibn-Qurra and al-Battani. His analysis of these measurements led him to conclude that al-Battani's value "agrees best with what we find by observation on the advance of the solstice in our own time."

The second stage was to find the relationship between the solar year and the mean lunar month, which is slightly more than 29.5 days. Gros-seteste, in his first treatise on calendar reform, *Canon in kalendarium*, had used a quadruple nineteen-year cycle of seventy-six years to compile a set of new-moon tables. In his third treatise, *Compotus correctorius*, he calcu-lated the error involved in this method and proposed a new and much more accurate cycle, using the Islamic year of twelve lunar months, or slightly more than 354 days. The cycle comprised thirty Islamic years, which together equal 10,631 days, the shortest period in which the cycle of lunations, or new moons, repeats itself. Grosseteste developed a method for combining this Islamic lunar cycle with the Christian solar calendar in order to calculate lunations accurately.

The third stage of Grosseteste's calendar reform employed his new method for an accurate reckoning of Easter. He writes in his *Compotus correctorius* that even without an accurate measurement of the solar year, the time of the spring equinox, on which the date of Easter depended, could be determined "by observations with instruments or from verified astronomical tables."

His treatise *On Prognostication* discusses astrological influences, along with his theory of tidal action, but he later condemned astrology, calling it a fraud and a delusion of Satan.

Grosseteste's most renowned disciple was Roger Bacon (c. 1220–1292), who acquired his interest in natural philosophy and mathematics while studying at Oxford. He received an master's either at Oxford or Paris, in around 1240, after which he lectured at the University of Paris on various works of Aristotle. He returned to Oxford in about 1247, when he met Grosseteste and became a member of his circle, which included Adam Marsh, Grosseteste's close friend and successor as lecturer at Oxford. Bacon writes of Grosseteste and Marsh in describing his intellectual development.

> There have been found some famous men, such as Robert, bishop of Lincoln, Brother Adam of Marsh, and many others, who have known how to utilize the power of mathematics to unfold the causes of all things and to give an explanation of human and Divine phenomena; and the assurance of this fact is to be found in the writings of these great men, as for instance in their works on the impressions [of the elements], on the rainbow, on comets, on the investigation of the places of the world, on celestial things, and on other questions appertaining both to theology and to natural philosophy.

It seems that Bacon became a Franciscan monk in around 1257, and soon afterward he experienced difficulties, probably because of a decree restricting the publication of works outside the order without prior approval. In any event, Pope Clement IV issued a papal mandate on June 22, 1266, asking Bacon for a copy of his philosophical writings. The mandate not only ordered Bacon to send his book but to state "what remedies you think should be applied in these matters which you recently intimated were of such great importance" and "to do this without delay as secretly as you can."

He eventually sent Clement three works—*Opus maius, Opus minus,* and *Opus tertium*—along with a letter proposing a reform of learning in the Church. He maintained that there were two types of experience, one obtained through mystical inspiration and the other through the senses, assisted by instruments and quantified in mathematics. The program of study that he recommended included languages, mathematics, optics, experimental science, and alchemy, followed by metaphysics and moral philosophy, which, under the guidance of theology, would lead to an understanding of nature and through that to knowledge of the Creator.

Within the next few years Bacon wrote three more works, the *Communia naturalium, Communia mathematica,* and *Compendium studii philosophie,* the last of which castigated the Franciscan and Dominican orders for their educational practices. At some time between 1277 and 1279 he was condemned and imprisoned in Paris by the Franciscans, possibly because of their censure of heretical Averroist ideas. Nothing further is known of his life until 1292, when he wrote his last work, the *Compendium studii theologii.*

He appropriated much of Grosseteste's "Metaphysics of Light" with its "multiplication of species," as well as his mentor's emphasis on mathematics, particularly geometry. In his *Opus maius* Bacon stated that "in the things of the world, as regards their efficient and generating causes, nothing can be known without the power of geometry," and he also said that "every multiplication is either according to lines, or angles or figures."

Bacon's ideas on optics repeat and extend those of Grosseteste, such as in his wave theory of light. Bacon held that light could propagate from one point to another only through a continuous medium and not through a vacuum.

Following Grosseteste, Bacon held that the "multiplication of species" proceeded as a series of pulses propagating through a medium both for light and sound, as he wrote in *Opus maius*:

> For sound is produced because parts of the object struck go out of their natural position, where there follows a trembling of the parts in every direction along with some rarefaction, because the motion of rarefaction is from the center to the circumference, and just as there is generated the first sound with the first tremor, so is there a second sound with the second tremor in a second portion of the air, and a third sound with a tremor in a third portion of the air, and so on.

He pointed out the differences between the propagation of sound, light, and odors:

> In the multiplication of sound a threefold temporal succession takes place, no one of which is present in the multiplication of light.... However the multiplication of both as regards itself is successive and requires time. Likewise in the case of odour the transmission is quite different from that of light, and yet the species of both will require time for transmission, for in odour there is a minute evaporation of vapour, which is, in fact, a body diffused in the air to the senses beside the species, which is similarly produced.... But in vision nothing is found except a succession of the multiplication.

He then went on to describe an observation showing that light propagates far more rapidly than sound:

> The fact that there is a difference in the transmission of light, sound, and odour can be set forth in another way, for light travels far more quickly in air than the other two. We note in the case of

one at a distance striking with a hammer or a staff that we can see the stroke delivered before we hear the strike produced. For we perceive with our vision a second stroke, before the sound of the first stroke reaches the hearing. The same is true of a flash of lightning, which we see before we see the sound of the thunder.

He went beyond Grosseteste in his explanation of vision, to which he paid particular attention because, as he said, "by means of it we search out certain experimental knowledge of all things that are in the heavens and earth." He gives a better description of the eye and optic nerves than any other Medieval Latin writer. Presenting an optical diagram of the human eye in the *Opus maius*, he noted, "I shall draw, therefore, a figure in which all of these matters are made as clear as possible, but a full demonstration would require a body fashioned like the eye in all the particulars aforesaid. The eye of a cow, pig or other animal can be used for illustration, if anyone wishes to experiment." As Crombie wrote of Bacon's work in this field: "His account of vision was one of the most important written during the Middle Ages and it became a point of departure for seventeenth-century work." Crombie goes on the quote Bacon in what he calls "a worthy expression of the ideals of the experimental method by one of its founders."

Bacon followed Grosseteste in suggesting the use of lenses as an aid to vision, which would soon lead to the invention of spectacles. He made a detailed study of the optics of vision and formulated a set of eight rules classifying the properties of convex and concave spherical surface in conjunction with the eye. As he wrote in the *Opus maius*:

> If anyone examines letters or other small objects through the medium of a crystal or glass or some other transparent body placed above the letters, and if it be shaped like the lesser segment of a sphere with the convex side towards the eye, and the eye is in the air, he will see the letters much better and they will appear much larger to him…. For this reason the instrument is useful to old people and people with weak eyes, for they can see any letter however small if magnified enough.

His scientific method was clearly stated in part 6 of the *Opus maius*, "De scientia experimentali," which also derives from Grosseteste.

> I now wish to unfold the principles of experimental science, since without experience nothing can be sufficiently known. For there are two modes of acquiring knowledge, namely by reasoning or experience. Reasoning draws a conclusion and makes us grant that conclusion, but does not make the conclusion certain, nor does it remove doubt so that the mind may rest on the intuition of truth, unless the mind may discover it by the world of experience.

He then wrote of the "three great prerogatives" of experimental science, the first being "that it investigates by experiment the noble conclusions of all the sciences." The second, according to Bacon, is that experiment adds new knowledge to existing sciences, and the third is that it creates entirely new areas of science. The new areas included, for example, those in astronomy that would be opened up by the telescope, and in medicine and biology by the microscope, two instruments whose discovery he predicted, along with applications of science in technology that he would write about in his *Epistola de secretis operibus artis et naturae et de nullitate magiae*. This describes marvelous machines such as self-powered ships, automobiles, airplanes, and submarines:

> Machines for navigation can be made without rowers so that the largest ships on rivers or seas will be moved by a single man in charge with greater velocity than if they were full of men. Also cars can be made so that without animals they will move with unbelievable rapidity.... Also flying machines can be constructed so that a man sits in the midst of the machine revolving some engine by which artificial wings are made to flap like a flying bird.... Also a machine can easily be made for walking in the sea and rivers, even to the bottom without danger.

The vital importance of mathematics in science is emphasized by Bacon, who wrote that "no science can be known without mathematics." He claimed that "in mathematics only are there the most convincing demonstrations through a necessary cause.... Wherefore it is evident that if, in the other sciences, we want to come to certitude without doubt and to truth without error, we must place the foundations of knowledge in mathematics." He went on to say that Grosseteste and Adam Marsh had followed this method and "if anyone should descend to the particular by applying the power of mathematics to the separate sciences, he would see that nothing magnificent in them can be known without mathematics."

Writing of the use of mathematics in making known "the things of this world," Bacon gave as an example astronomy, which "considers the quantity of all things that are included among the celestial and all things that are reduced to quantity." He said that "by instruments suitable to them and by tables and canons," one can measure the movements of the celestial bodies and reduce them to rules on which predictions can be based. He also employed Grosseteste's *Compotus* to carry on his program of calendar reform, making use of both Aristotle's physical model of the concentric celestial spheres and Ptolemy's mathematical method of eccentrics and epicycles.

Bacon also used his scientific method to study the rainbow. He began by examining phenomena similar to rainbows, including the dispersion of colors in crystals, morning dew on grass, water spray, and light refracted by a glass vessel filled with water or seen through cloth or partially closed eyelashes. He then examined the rainbow itself, observing that it always appeared when there was a cloud or mist; that the bow was always opposite the sun, that the center of the bow was always in a straight line with the observer and the sun, and that there was a definite relation between the altitude of the rainbow and the sun. Reporting on his observation of a rainbow with an astrolabe, he noted: "The experimenter, therefore, taking the altitude of the sun and of the rainbow above the horizon, will find that the final altitude at which the rainbow can appear above the horizon is 42°, and this is the maximum elevation of the rainbow.... And the rainbow reaches this maximum elevation when the sun is on the horizon, namely at sunrise and sunset."

Grosseteste's theory of the rainbow was improved by Bacon, who realized that the phenomenon was due to the action of individual raindrops, though he erred in rejecting refraction as part of the process. He was also mistaken in his explanation of the colors of the rainbow, which he thought were illusory. He concluded that "the bow appears only in raindrops from which there is reflection to the eye; because there is merely the appearance of colours arising from the imagination and deception of the vision.... A reflection comes from every drop at the same time, while the eye is in one position, because of the equality of the angles of incidence and reflection."

Although Bacon's theory of the rainbow was incorrect, his observations and method, following and extending those of Grosseteste, paved the way for those who would eventually arrive at the true explanation. This was finally achieved by Newton in his *Optiks*, where he explained the rainbow as being due to a combination of reflection, refraction, and dispersion, or the division of sunlight into its component colors, a spectrum extending from red to violet.

Bacon's tendency toward the occult is evidenced in a number of his statements, for example, where writes that "it has been proved by certain experiments" that life can be greatly extended by "secret experiences." One of his recommendations for achieving an exceptionally long life involves eating the specially prepared flesh of flying dragons, which he says also "inspires the intellect," or so he was told "without deceit or doubt from men of proved trustworthiness." Writings such as this gave Bacon the posthumous reputation of being a magician and diviner who had learned his black arts from Satan.

But this should not obscure the fact that he and Grosseteste were among the founders of the experimental method that gave rise to modern science. Bacon's account of his method in the *Opus maius* is an expression of his plea, addressed to Pope Clement IV, for the study of science, one that would be answered by those who followed the path that he and Grosseteste had opened.

The crucial advance made by Grosseteste and his followers beyond their immediate predecessors was explained succinctly by Crombie: "The strategic act by which Grosseteste and his thirteenth- and fourteenth-century successors created modern experimental science was to unite the experimental habit of the practical arts with the rationalization of twelfth-century philosophy." Although Grosseteste himself did not always adhere to the practice of verifying his theories by experimentation, most of his followers did. As Crombie noted:

> In the next generation such natural philosophers as Roger Bacon and Petrus Peregrinus and, later, Theodoric of Freiburg were to use this principle as the basis of some really thorough and elegant pieces of experimental research.
>
> Hence reasoning does not attest these matters, but experiments on a large scale made with instruments and by various necessary means are required. Therefore no discussion can give an adequate explanation in these matters, for the whole subject is dependent on experiment. For this reason I do not think that in this matter I have grasped the whole truth, because I have not yet made all the experiments that are necessary. . . . Therefore it does devolve on me to give at this time an attestation impossible for me, but to treat the subject in the form of a plea for the study of science.

7

The Experimental Method

NE OF THE EARLIEST ADOPTERS OF THE EXPERIMENTAL METHOD of Robert Grosseteste and Roger Bacon was Peter Peregrinus. All that is known of him comes from his work and from references by Bacon, who calls him *"magister Petrus . . . dominus experimentorum."* Bacon, writing in 1267 in his *Opus tertium*, gave this account of the experimental method followed by Peter:

> And this science certifies all natural and artificial things in the particular and in the proper discipline by perfect experiment; not by argument, like the purely theoretical sciences, nor by weak and imperfect experiences, like the practical sciences. And therefore this science is the master of all the preceding sciences, and the end of all theoretical argument.... One man I know [i.e., Peter], and only one, who can be praised for his achievement in this science.

Bacon went on to describe Peter's manifold activities in the practical application of science:

> Through experiment he gains knowledge of natural things, medical, chemical, and indeed of everything in the heavens and earth. He is ashamed that things should be known to laymen, old women, soldiers, ploughmen, of which he is ignorant. Therefore he has looked closely into the doings of those who work in metals and minerals of all kinds; he knows everything relating to the art of war, the making

of weapons, and the chase; he has looked closely into agriculture and farming work.... For the past three years he has been working at the production of a mirror that shall produce combustion at a fixed distance; a problem that the Latins have neither solved nor attempted, though books have been written on the subject.

The only extant work by Peter Peregrinus is his *De magnete*, a treatise on magnetism. This is actually a letter, the *Epistola Petri Peregrini de Maricourt ad Sygerum de Foucaucourt, miltem, de magnete* (*Letter on the Magnet of Peter Peregrinus of Maricourt to Sygerus of Foucaucourt, Soldier*). Peter concluded the letter with the note that it had been "completed in camp, at the siege of Lucera, in the year of our Lord 1269, eighth day of August." This would indicate that Peter was at the time in the army of Charles of Anjou, king of Sicily, who was then besieging the city of Lucera in southern Italy.

Peter's statement of purpose in book 1 of *De magnete* reveals that he was following the experimental method of Robert Grosseteste and Roger Bacon in his investigation of magnetism:

> You must realize, dearest friend, that the investigator in this subject must understand nature and not be ignorant of the celestial motions, he must also be very diligent in the use of his own hands, so that through the operation of this stone [i.e., the lodestone, or magnetic ore] he may show wonderful effects. For by his industry he will then in a short time be able to correct an error which he would never do in eternity for his knowledge of natural philosophy and mathematics alone, if he lacked carefulness with his hands. For in investigating the unknown we greatly need manual industry, without which we can usually accomplish nothing perfectly. Yet there are many other things subject to the rule of reason which we cannot investigate completely by the hand.

De magnete is in two parts, of which the first is an account of the experiments that Peter performed in his investigation of magnetism, and the second is a description of instruments that he made as the results of his studies. As Peter wrote in the first chapter of book 1, after a summary of his treatise: "We shall not communicate in this epistle any information save about the manifest properties of this stone, on the ground that this teaching will form part of a tract in which we shall show how to construct physical instruments."

Peter began book 1 by explaining how to recognize a lodestone, particularly by its attraction to iron. Taking a spherically shaped lodestone, later to be called a *terrella*, he used an iron needle as an indicator and placed it at various points on the sphere to show that the direction of the magnetic force, marking it on the stone, as he noted: "Let a needle or elongated piece of iron, slender like a needle, be placed on the stone, and a line be drawn along the length of iron dividing the stone in the middle. Then let the needle or iron be placed in another position on the stone and mark the stone in a similar manner according to that position."

The lines of magnetic force, so drawn, form meridians that converge at the two poles of the sphere, just like the meridian geographical lines on the globe of the earth or those of the celestial sphere. The poles can also be found by marking the two points where the needle clings to the sphere with the greatest force, or where a short needle will stand upright.

Peter then described a demonstration to distinguish between the north and south magnetic poles of a lodestone. If the loadstone is placed on a float placed in a vessel filled with water, so that "the stone may be like a sailor on a ship, i.e., free to turn in any direction, then that it will align itself" so that "each part of the stone is in the direction of its own part of the heavens." That is, the north pole of the magnet will point toward the direction of the north celestial pole and the south magnetic pole will face the south celestial pole. I have done this same demonstration in my elementary physics classes, and it is replicated in primary education throughout the world.

Using another magnet as a probe, Peter found that that the south pole of the floating magnet always followed the north pole of the other magnet when it was brought nearer, and vice versa. He observed that if like poles of the two magnets were brought close together, "the stone which you hold in your hand will appear to flee the floating stone."

He discovered that if an oblong piece of iron was touched by a magnet, it became magnetized, with its north and south poles oriented oppositely to the original magnet. But the polarity of the newly magnetized piece of iron could be reversed by "violence" with a lodestone. Peter said that "the cause of this is the impression of that which acted last, confounding and altering the virtue of the first."

Peter found that if he broke a lodestone into two or more pieces, each with a north and south pole, then these pieces were cemented together in the same alignment they had originally, they formed a single magnet with the same polarity it had before. This experiment, which is also still done in elementary science classes, led Peter to conclude that "the agent strives not only to join its patient to itself but to unite with it."

He then wrote of his efforts to determine "whence the magnet receives the natural virtue which it has." Such virtue does not come just from deposits of lodestone at the north geographic pole of the earth, according to Peter, because loadstone is mined in many parts of the globe. Besides, he said, since magnets point south as well as north, "we are right in supposing that the virtue in the poles of the stone flows in not only from the northern part but also from the southern part." Furthermore, the north pole of a magnet does not point in the direction aligned exactly with Polaris, the pole star, but to the meeting point of the geographic meridians of the earth. He knew that Polaris does not coincide exactly with the north celestial pole but rotates in a small circle around it. He was unaware, though, of the precession of the equinoxes, which causes the celestial pole to slowly precess in a circle about the perpendicular to the ecliptic. Moreover, the earth itself is a huge magnet, with its north and south magnetic poles close to its north and south geographic poles. Unaware of this, Peter believed that every part of a spherical magnet received its power from the corresponding part of the celestial sphere, concluding that "from these facts therefore it is manifest that it is from the poles of the heavens that the poles of the magnet receive their virtue." He said that this could be demonstrated by fixing a spherical lodestone between two sharp pivots so that it could move like an armillary sphere, in which case it would rotate and follow the movement of the heavens.

Peter described two instruments that combine an astrolabe and a magnet, designed to measure the azimuth of celestial bodies. One of them used a floating magnet and in the other the magnet was fixed between metal pivots, so that in both cases it could rotate to follow the motion of the heavenly bodies. By measuring the azimuth with these instruments time could be told, Peter said, and when longitudes and latitudes were known one could navigate on land and sea. This was the first attempt to invent a celestial navigation device, though it was based on the false premise that the earth's magnetic and geographic poles are identical.

Peter also invented what he hoped would be a perpetual motion machine, which, he said, "will move continually and perpetually" by the action of a fixed loadstone upon small iron magnets attached to the periphery of a wheel. When the machine did not move perpetually, he blamed the failure on his lack of skill rather than the impossibility of creating an eternal source of energy. This was the first of many attempts to create a perpetual motion machine and would eventually lead to the law of conservation of energy, one of the most basic principles of physics. The law shows that perpetual motion machines are impossible because of the inevitable loss of energy through friction, air resistance, and other factors.

De magnete was very popular in the late medieval era, as evidenced by the fact that there are at least thirty-one extant manuscript copies. No further study of magnetism on this scale and level was made until 1600, when William Gilbert published his work of the same title, acknowledging his debt to Peter Peregrinus and incorporating all of his demonstrations in his own thesis.

Another pioneer of the new European science was Jordanus Nemorarius (fl. c. 1220), a contemporary of Grosseteste. Virtually nothing can be said about the life of Jordanus, who is known only through the inclusion of his works in the *Biblionomia*, a catalogue of the library of Richard de Fournival done sometime between 1246 and 1260, in which twelve treatises are ascribed to him.

Jordanus made his greatest contribution in the medieval "science of weights" (*scientia de ponderibus*), now known as statics, the study of forces in equilibrium. One of the concepts he introduced was that of "positional gravity" (*gravitas secundum situm*), which he expressed in the statement that "weight is heavier positionally, when, at a given position, its path of descent is less oblique" and that "a more oblique descent is one in which, for a given distance, there is a smaller component of the vertical." An example would be a block on an inclined plane, whose apparent weight, the force with which it presses against the surface, is greater if the angle of inclination is less. This is equivalent to resolving the weight into two components, one perpendicular to the plane, which is the apparent weight or "positional gravity," and the other parallel to the surface.

Jordanus applied the concept of positional gravity in his study of the most basic problem in statics, that of the beam balance, where two weights are suspended on either side of the fulcrum. His experiment demonstrated that "when the beam of a balance of equal arms is in horizontal position, then if equal weights are suspended from its extremities, it will not leave the horizontal position; and if it should be moved from the horizontal position, it will revert to it." He then went on to explain that "if the arms of the balance are unequal, then if equal weights are suspended from their extremities, the balance will be depressed on the side of the longer arm." Here Jordanus coined the term *positional gravity*, which in this case is equal to the weight of the object times its lever arm, the perpendicular distance from the fulcrum to the line of action of the weight. This is now known as the moment of a force, or torque, a measure of its effectiveness in rotating the balance, equilibrium resulting if the two torques are equal and opposite.

Jordanus then demonstrated the law of the lever, as is still done in elementary physics classes, showing that two objects will balance each other if their weights are inversely proportional to their lever arms. Here he used the concept of "work," the product of the weight of an object times the distance through which it is lifted or otherwise moved, the first clear definition of this fundamental concept in physics. He also introduced the concept of "virtual velocity," that is, one that is infinitesimally small, since a real movement cannot take place in a system under equilibrium. For example, when two objects balanced on a lever, the positive work done in lifting one weight is equal to the negative work in lowering the other, leading to the conclusion that the system is in equilibrium. His proof makes use of what is known as the "axiom of Jordanus," that the motive power that can lift a given weight a certain height can lift a weight k times heavier to $1/k$ times that height, where k is any number.

The same concepts were applied in studying the equilibrium of two different connected weights on inclined planes of different inclinations, which Jordanus treated as a generalized case of the law of the lever. The proof refers to a triangle ABC, which has BC as its base and a right angle at A, where a pulley connects two weights, with w(1) on side AB and w(2) on side AC. He showed that the two weights will be in equilibrium if their positional gravities are equal, that is, if the components of each weight down its plane is equal to that of the other in the opposite direction. This can be reduced to the equation w(1)/w(2) = AB/AC, which is equivalent to the law of the lever.

Jordanus also did research in mathematics, and he is credited with treatises on geometry, algebra, proportions, and both theoretical and practical arithmetic, as well as a work on mathematical astronomy.

His four-book treatise *Liber philolotegni de triangulis* has been described by historian Edward Grant as "medieval geometry at its highest level." The treatise contains propositions concerning the ratios of sides and angles; the divisions of straight lines, triangles, and quadrangles under a variety of given conditions; and with ratios of arc length and plane segments in the same circle as well as in different circles. Here Jordanus, following Archimedes, solved problems involving the determination of the center of gravity of triangles and other plane figures. He took his solutions from Latin translations of Graeco-Islamic works, including one proposition from Alhazen's *Optics*. One proof on the quadrature of the circle is not found in Archimedes and may be an original contribution by Jordanus.

Another treatise focusing on algebra, the treatise *De numeris datis* is more Euclidian in its approach than other Latin works derived from Arabic writers. It was praised by the great fifteenth-century mathematician and astronomer Regiomontanus, according to mathematics historian Carl B. Boyer, who also says that it anticipates the pioneering work of the sixteenth-century mathematician François Viète in the application of analysis to algebraic problems. In this work Jordanus was the first to use the letters of the alphabet in arithmetical problems for greater generality, and he presented algebraic problems leading to linear and quadratic equations.

These other writings were profoundly influential, but the best-known mathematical work of Jordanus is the *Arithmetica*, which became the standard source of theoretical arithmetic in the Middle Ages. Comprising more than four hundred propositions in ten books, *Arithmetica* is modeled on the arithmetical books in Euclid's *Elements*. Nevertheless, the proofs frequently differ from those in Euclid. A typical proposition in the *Arithmetica*, which has no counterpart in the *Elements*, is the ninth in book 1: "The result of the multiplication of any number by however many numbers you please is equal to the result of the multiplication of the same number by the number composed of all the others." That is to say, if $A \cdot B = D$ and $A \cdot C = E$, then $D + E = A(B + C)$.

Jordanus also wrote a treatise called *Liber de proportionibus*, which contains propositions similar to those in book 5 of Euclid's *Elements*. A treatise called *Demonstratio de minutiis*, an algorithm of fractions, may also be by Jordanus. This describes arithmetic operations with fractions alone as well as with fractions and integers.

Still another mathematical work of Jordanus is the *Suppletiones plane sphera*, which is probably a commentary on Ptolemy's *Planisphaerium*. Historian George Sarton described this as "a treatise on mathematical astronomy, which contains the first general demonstration of the fundamental property of stereographic projection—i.e., that circles are projected only as circles (Ptolemy had proved it only in special cases)."

The treatise *Demonstratio Jordani algorismo*, which is attributed to Jordanus, is one of the earliest Latin works mentioning the Arabic number system. This work differed radically from Sacrobosco's *Algorismus vulgaris* (*Common Algorism*) in that Jordanus described the arithmetical operations briefly and formally without giving examples.

The new scientific methodology that had been introduced by Robert Grosseteste was further developed by John Duns Scotus (c. 1260–1308) and William of Ockham (c. 1285–1349). Both of them were Franciscans who followed Grosseteste's lead in their methodology, studying phenomena by collecting similar examples and using the principle of uniformity and economy in order to achieve certitude in knowledge of the things of experience.

Scotus differed with Grosseteste in absolutely rejecting the idea that man could not know anything with certainty except through divine illumination. Although complete knowledge of universals was not possible without a complete study of all particulars, he said, nevertheless probable knowledge could be obtained by induction by studying a large enough sample of particular cases. He believed that certain knowledge was possible of three types of "knowable things," first of self-evident basic principles, such as that the (finite) whole is greater than the part; second, sensory experience, although judgments about observations might be false; and third, consciousness of personal actions and states of mind.

The three types of certainty posited by Scotus were accepted by Ockham, who added two additional principles. The first was the principle of evidence, in which he distinguished between the science of real entities, those that were known by experience to exist and whose names stood for things existing in nature, and the science of logical entities, which dealt with logical constructions whose names merely represented concepts. As philosopher and medievalist Ernest A. Moody noted, "Ockham holds that our knowledge of things is based on a direct and immediate awareness of what is present to our senses and intellect, which he calls intuitive cognition." He believed that only intuitive knowledge obtained through experience of individual things could give certain knowledge of the real. His second principle, which came to be known as "Ockham's razor," was that of economy in explanation, usually expressed in the statement "What can be accounted for by fewer assumptions is explained in vain by more." Ockham himself stated it in a longer form: "Nothing is to be assumed as evident, unless it is known *per se*, or is proved by the authority of Scripture."

Ockham held that it was possible to define logical rules for establishing causal connections, later to be known as the method of agreement and difference, which he describe thus:

> This is sufficient for anything being an immediate cause, namely that when it is present the effect follows and when not present, all other conditions being the same, the effect does not follow.... That this is sufficient for anything to be an immediate cause of anything else is clear because there is no way of knowing that something is an immediate cause of something else.... All causes properly so-called are immediate causes.

Thus, as Crombie wrote, "The practical programme for natural science was simply to correlate observed facts," or "save the appearances," by means of logic and mathematics.

This method was used by Ockham to deal with the problem of motion, such as that of a projectile fired by a catapult. According to Aristotle, every cause has an effect and the effect ends when the cause is removed. As noted earlier, Aristotle explained the continued motion of a projectile after it leaves the catapult as being due to a process he called antiperistasis, in which the air in front of the moving body streams around behind it in a way that gives it a push forward. Ockham rejected this notion as well as the theory of impetus, which had first been proposed by John Philoponus. Instead he held that motion is an abstract concept, having no reality aside from moving bodies, which was used to describe the fact that an object changed its spatial relationship with some other body. There was no need to postulate any external or internal efficient cause, such as antiperistasis or impetus, to explain such a sequence, as he explained, using Ockham's razor:

Motion is not such a thing wholly distinct from the permanent body, because it is futile to use more entities when it is possible to use fewer. But without any such thing we can save the motion and everything that is said about motion. Therefore it is futile to postulate such other things. That without such an additional thing we can save motion and everything that is said about it is made clear by considering the separate parts of motion. For it is clear that local motion is to be conceived of as follows: positing that the body is in one place and later in another place, thus proceeding without any rest or any intermediate thing other than the body itself and the agent itself which moves, we have local motion truly. Therefore it is futile to postulate such other things.

This idea was then applied by Ockham to deal with the question of projectile motion, thus forming a conception that was to be used in the seventeenth-century theory of inertia, one of the basic principles of the new dynamics that would be formulated by Newton.

I say therefore that the moving thing in such a motion [i.e., projectile motion], after the separation of the moving body from the first projector, is the moved thing itself, not by reason of any power in it; for this moving thing and the moved thing cannot be distinguished. If you say that a new effect has some cause and local motion is a new effect, I say that local motion is not a new effect ... because it is nothing else but the fact that the moving body is in different parts of space in such a manner that it is not in any one part, since two contradictions are not both true.

This method was also used by Ockham in dealing with an Averroist theory of magnetism that became popular with French philosophers of nature in the late thirteenth and early fourteenth centuries. This theory purported that a magnet did not really attract a piece of iron with a force acting at a distance, but induced in it a "magnetic quality" or "virtue," giving it the power to move itself. Thus the apparent attractive force of the magnet was actually due to a moving power within the power itself, according to the theory. A version of this theory, based on Grosseteste's concept of multiplication of species, was proposed by the French Averroist Jean de Jandun (d. 1328) in his commentary on a work of Aristotle. Jandun suggested that the power or "virtue" that moved the iron was due to a *species magnetica* that had been induced in it, having been propagated through the medium point-to-point by successive modification of parts in contact until it reached the iron and modified it.

Ockham applied his principle of economy to these theories, eliminating the hypothetical *species magnetica* that had been introduced as an intermediary simply to avoid postulating action at a distance. He thought that a force like that of the magnet on iron could act at a distance, just as the sun gave its light to the earth across the intervening space, and he rejected the Aristotelian notion that a moving power must be in intimate contact with the body moved. This was a revolutionary step toward some of the most important developments leading to the new science that emerged from the seventeenth century onward. Crombie wrote of how Ockham's razor allowed later philosophers of nature to "save the appearances" in the Greek sense, that is, to account for physical phenomena, such as action at a distance, without superfluous hypotheses.

> He could see no reason to object to action at a distance, and he cut out from his explanation the intermediate "species" which were postulated simply to avoid such action at a distance and were not necessary to "save the phenomena." ... Ockham's arguments prepared the way for Newton to "save the appearances" by his theory of universal gravitation, and it is interesting that in the seventeenth century Leibniz brought the same philosophical objection to action at a distance against this theory. In the study of magnetism precisely the same philosophical problem led Faraday to propose his lines or tubes of force; Michell and Coulomb, inspired by Newton, had shown how to save "the appearances" with the law that the strength

of the magnetic attraction was inversely proportional to the square of the distance.

Thus with one stroke Ockham's razor cut away the intellectual obstacles that for centuries had impeded an understanding of the macroscopic forces in nature, gravitational and electromagnetic, clearing the path for Newton and those who followed him in the development of modern science.

Such was the complicated development of the scientific method during the thirteenth and fourteenth centuries, with its debates on the complex relation of philosophical principles, empirical observations, directed experiments, and mathematical theories, which in the sixteenth and seventeenth centuries would sweep away the ancient Aristotelian system in the emergence of modern science. A. C. Crombie wrote of the centrality of Aristotle throughout this period. He said, "The purely theoretical criticism of Aristotle's theory of science and fundamental principle ... was to lead later to the overthrow of the whole system of Aristotelian physics:

Much of it developed from within Aristotle's thought itself. Indeed Aristotle can be seen as a sort of tragic hero striding through medieval science. From Grosseteste to Galileo he occupied the centre of the stage, seducing men's minds by the magical promise of his concept, exciting their passions and dividing their allegiances. In the end he forced them to turn against him as the clear consequences of his undertaking gradually became clear; and yet, from the depths of his own system, he provided many of the weapons with which he was attacked.

He went on to describe how the new scientific method, with its balanced combination of philosophy, directed experiment, and mathematical theory changed the whole approach to scientific inquiry.

The most important of these weapons were made by new ideas on scientific method, especially by new ideas on induction and experiment, and on the role of mathematics in explaining physical phenomena. These gradually led to an entirely different conception of the kind of question that should be asked in natural science, the kind of question, in fact, to which the experimental and mathematical methods could give an answer.

As we will see, the area where this new method would prove crucial was dynamics, the study of the relation between force and motion, and this was where Aristotle's ideas in this field were questioned on the most fundamental level, on the very nature of motion as a phenomenon occurring in time and space.

8

The Science of Motion

*D*URING THE SECOND QUARTER OF THE FOURTEENTH CENTURY A
group of scholars at Merton College, Oxford—Thomas Bradwardine, William Heytesbury, John of Dumbleton, and Richard
Swineshead—developed the conceptual framework and technical vocabulary of the new science of motion. Their work continued the Oxford
tradition in science initiated by Robert Grosseteste and Roger Bacon. The
fundamental difference between the approach of the Merton scholars and
that of Grosseteste and Bacon is noted by Crombie:

> In the 13th century it was mainly the philosophical issues that de-
> termined the terms of the discussion of motion, but this gave way
> in the 14th century to a greater attention to the mathematical and
> quantitative formulation of laws of motion. Attention began to turn
> from the "why" to the "how." Practically without exception—the
> most significant was William of Ockham—the natural philosophers
> of this period based their discussions on the accepted Aristotelian
> principle that being in motion meant being moved by something.
> Differences of opinion concerned the nature of the moving power
> in the different cases and the quantitative relations between the dif-
> ferent determinants of velocity.

As we shall see, these discussions on dynamics would lead to the development of kinematics, the purely mathematical description of motion,

anticipating the work of Galileo, as well as a revival of the impetus theory of John Philoponus, which would eventually lead to Newton's second law of motion. The discussions would also bring up the suggestion that the sun is the center of the cosmos, and not the earth, two centuries before the Copernican theory, and also the idea that the universe is infinite.

Thomas Bradwardine (c. 1290–1349) received his bachelor's, master's, and doctoral degrees at Oxford in the years 1321–1348, and from 1323 until 1335 he was a fellow of Merton College. In 1339 he was chaplain and perhaps confessor to King Edward III and accompanied the king to France in the campaign of 1346. He was elected archbishop of Canterbury on June 4, 1349, but he died of the plague on August 26 of that same year.

According to Crombie, Bradwardine "was the real founder of the school of scientific thought associated with Merton College, and he stood in the same relationship to the Oxford thought of the fourteenth century as Grosseteste did to that of the thirteenth."

In his principal work, the *Tractatus de proportionum*, completed in 1328, Bradwardine gives a clear mathematical account of the laws of motion so as to give the relationship among force, resistance, and velocity, expressing his formulas verbally in what has been called "word algebra" since he had no mathematical notation at his disposal.

Bradwardine was influenced by the scholar Gerard of Brussels, who appears to have been associated with Jordanus. Gerard seems to have been the first European to deal with kinematics, a purely mathematical description of motion. His treatise on kinematics *De motu*, written at some time between 1187 and 1260, was strongly influenced by Euclid and Archimedes, making use of the latter's characteristic proof by reductio ad absurdum and his method of exhaustion. Gerard's work on motion was part of his general study of what was known as "the intension and remission of forms," the variation in intensity of a quality or essence. This dealt with the problem of how to describe a subject in which the intensity of a quality varies from one point to another. An analogous question was how to name and measure a motion in which the velocity varies from one moment in time to another or from point to point in the moving body, such as in a rotating wheel. Crombie gave a succinct summary of Gerard's approach and its influence on Bradwardine and his successors at Oxford:

> Dealing with movements of rotation, Gerard took an approach that became characteristic of modern kinematics, seeing as the basic objective of analysis the representation of non-uniform velocities by uniform velocities. Although he fell short of defining velocity as a ratio of unlike qualities [i.e., distance and time], he seems to have assumed that the speed of a motion can be assigned some number or quantity making it a magnitude like space or time. Bradwardine specifically discussed some of Gerard's propositions, and it seems probable that *De motu* directed the attention of the Oxford mathematicians of the 14th century to the kinematic description of variable motion and to the metric definition of velocity required for their treatment.

H. Lamar Crosby published an edition of the *Tractatus de proportionibus* in 1955 in which he paraphrased Bradwardine and expressed his formulations in modern notation. For example, as Crombie quotes Crosby: "The equation that he asserted held for all proportions greater than 1 of power (m) to resistance (r) can be expressed in modern terminology, where v is velocity, as follows: $v = \log(m/r)$." Crosby shows how Bradwardine first disproved several widely held views about the laws of motion and then went on to formulate his own version.

The law of motion held by most medieval Aristotelians, expressed in modern form, states that the velocity (v) of an object is directly proportional to the motive force (F) divided by the resistance (r) of the medium. If this law was valid, then, as Crosby paraphrases Bradwardine, "The first false consequent is that any force, however small, can move any resistance, however large," which he dismisses with the statement that "this theory is seen to be a violation of the universally accepted Aristotelian notion that there should be a finite velocity for any finite values of force and resistance." He then says that "the second false consequent is similar," because if this law is true then "any *mobile* [movable thing] can be moved by any force."

Two examples from experience are given by Bradwardine showing that velocity is not directly proportional to the ratio of force to resistance: "Furthermore, sense experience teaches the opposite of this view, for we see, for example, that if one man can scarcely move a heavy rock, then two men working together can move it more than twice as rapidly. The same principle is illustrated in the case of clock weights: to double a weight may more than double the speed of descent."

He focused on the change in velocity, rather than the velocity itself, and tried to show how this was related to force and resistance. Stated in modern mathematical terms, he looked for a functional relation between the dependent variable v and the two independent variables F and r; that is, given values for F and r to find the corresponding value of v. As noted above, after trying and rejecting a number of equations he finally settled on a law of motion, which, in modern terminology, states that the velocity is proportional to the logarithm of F/r.

When discussing his law of motion, Bradwardine makes a distinction of fundamental importance between what in modern physics are termed *average velocity* and *instantaneous velocity*. Average velocity is the ratio of distance traveled by a body to the time elapsed, whereas instantaneous velocity is the ratio when the time interval is reduced to an infinitesimal, that is, becomes vanishingly small. The values of the average and instantaneous velocities are the same if the body is moving with constant speed. Bradwardine's law of motion is primarily dynamic, and so it must be understood that his relation refers to the instantaneous velocity. Historian Richard C. Dales wrote of the significance of Bradwardine's concept of instantaneous velocity:

> This use of infinitesimals to express an instantaneous velocity considered as a quantity of motion, and the unanswered questions thus raised concerning the relationship between this quality of motion and the quantity of motion possessing this quality, were enormously

fruitful in subsequent studies of motion from the fourteenth through seventeenth century.

Bradwardine never tested his law of motion, which would have shown him that it is not correct. The reason for this is undoubtedly the difficulty that a scientist at that time would have had in measuring dynamical quantities, for even in my own elementary physics classes I have found it quite difficult to check the dynamical theories of medieval physicists using the equipment that would have been available to them, not to mention the fact that the logarithmic function had not yet been invented.

Nevertheless, his formulation of the problem in terms of a mathematical functional relationship was an important step forward in the science of dynamics, one that was followed by his successors at Oxford. Their researches established the foundations of the late medieval tradition of the *calculatores*, those who studied the quantitative variation of motion, power, and qualities in space and time. The tradition spread to the Continent in the second half of the fourteenth century; it was revived in Italy in the fifteenth century and again in Paris as well as Spain in the first third of the sixteenth century.

William Heytesbury's name, variously spelled, appears in the records of Merton College for 1330 and 1338–1339, and he may be the William Heighterbury or Hetisbury who was chancellor of the university in 1371. His most influential scientific work is his *Regulae solvendi sophismata*, dating from 1335.

The last chapter of the *Regulae* deals with the quantitative description of motion or change in the three Aristotelian categories of place, quantity, and quality. Each of the three subchapters tries to establish the proper measure of velocity in the given category.

Heytesbury followed what Gerard of Brussels had called the study of the "intension and remission of forms," trying to give quantitative expression to changes of quality, which at Merton College in Oxford was known as "the latitude of forms." The purpose of the method developed by Heytesbury and his followers was to give a numerical measure of the amount by which a quality or "form" increased or decreased according to some given scale. A "form" was defined as any variable physical quantity or quality, and the "extensio," that is, the "extension" or "longitude" of the form, was the numerical value that was assigned to it. Latitudes were categorized as "uniform" if they were of constant "*intensio*," or intensity, such as motion with constant velocity, or "difform" if the intensity varied in space or time, as in the case of accelerated or retarded motion. A "difform" change was said to be "uniformly difform" if the intensity changed linearly in space or time, as in cases where the acceleration or retarding is increasing or decreasing at a constant rate; otherwise it was called "difformly difform."

Having made these distinctions, Heytesbury then formulated a rule, which states that every uniformly difform latitude corresponds to its mean degree. Thus if the measure of the intensity of a quality at one end of a body is 2° and 4° at the other end, then the latitude of that quality is 3°. Heytesbury and his followers held—though without empirical verification— that intensities of a quality are numerically additive. Applying this to local motion, the intensity of which is measured in terms of distances, which are additive, covered in unit time, the rule leads to verifiable empirical results. As Heytesbury explains what came to be called the Merton mean-speed rule:

> For whence it commences from zero degrees or from some [finite] degree, every latitude [of velocity], provided that it is terminated at some finite degree, and is acquired or lost uniformly, will correspond to its mean degree. Thus the moving body acquiring or losing this latitude during some period of time, will traverse a distance exactly equal to what it would traverse in an equal period of time if it were moved continuously at its mean degree. For of every such latitude commencing from rest and terminating at some [finite] degree [of velocity], the mean degree is one-half the terminal degree of that same quality.

Applying this system to local motion, Heytesbury defined uniform acceleration as motion in which the velocity is changing at a constant rate, either increasing or decreasing. For such motion he defined acceleration as the change in velocity in a given time, which would be negative in the case of deceleration. He followed Thomas Bradwardine in using the notion of instantaneous velocity, that is, the speed at a particular moment, defining it as the distance traveled by a body in a given time if it continued to move with the speed that it had at that moment. Using the mean-speed rule, he showed that for uniformly accelerated motion the average velocity during a time interval is equal to the instantaneous velocity at the midpoint of that interval.

John of Dumbleton (fl. 1331–1349) is mentioned as a fellow of Merton College in the years 1338–1348 and is best known for his *Summa logicae et philosophiae naturalis*, a vast work that critically discusses most of the topics in physics and philosophy of his time. An important part of the discussion involved rates of change, including motion, change of quality, and growth, in relation to a fixed scale such as distance or time.

Building on the work of his Oxford colleagues, Dumbleton was one of the first to express functional relationships in graphical form, with the "latitude" or intensity on the vertical axis and the "longitude" or extension on the horizontal axis. If velocity (i.e., intensity of motion) is represented on the vertical axis versus time (i.e., extension of time), then uniform velocity would plot as a horizontal straight line, that is, the same velocity at all times. Uniformly difform motion, that is, motion with velocity changing uniformly in time, would plot as a straight line inclined to the horizontal axis, sloping upward if the acceleration is positive or downward if it is negative.

Dumbleton also gave an arithmetic proof of the Merton mean-speed rule, which he called the "method of latitudes," stating that "the latitude of a uniformly difform movement corresponds to the degree of the midpoint." He used this method in the *Summa* to study the problem of the variation in the strength of light as a function of the distance from its source. Using the sun or a candle as an example of a light source, he postulated a direct relationship between the strength of a luminous body and the distance through which it acted at a given intensity, and an inverse relationship between the density of the medium and the distance of action.

He concluded: "If, therefore, A acts weakly only because of the distance, then, according as they are proportionately distant, so A proportionately brings about degrees in them of the same degree which it has itself." He realized that the decrease in intensity of illumination was not linearly proportional to the distance, for he noted that "a luminous body does not act uniformly difformly in a uniformly dense medium." But he did not succeed in finding the exact quantitative relationship, which is that the intensity of illumination due to a luminous source is inversely proportional to the square of the distance, a law discovered by Johannes Kepler in 1604.

Richard Swineshead, who was known as the Calculator among his peers, was also a fixture at Merton College in the 1340s. He is best known as the author of the *Liber calculationum* (c. 1340–1350), a work that became famous for its extensive use of mathematics in physics, anticipating the approach of Newton. He may also have written two works on motion entitled *De motu* and *De motu locali*, as well as a partial commentary on Aristotle's *De caelo*.

The *Liber calculationum* concentrates on calculating the values of physical variables and solving problems about their changes. The work is divided into sixteen treatises, of which the last three are devoted to "Local Motion," where he elaborates at exhausting length on Bradwardine's law of dynamics in every conceivable type of motion, many of which have no known parallel in nature. The *Liber calculationum* was disseminated widely in Europe, and it was printed in Padua (c. 1477), Pavia (1498), and Venice (1520). The latter edition was later transcribed for the great German mathematician and philosopher Leibniz (1646–1716), who praised Swineshead for having introduced mathematics into scholastic philosophy.

Advances in the theory of motion were also being made in Paris, beginning with the work of Jean Buridan (c. 1295–c. 1358), who was deeply influenced by the ideas of Bradwardine and his followers at Oxford. Little is known of Buridan's origins other than the fact that he was born in the diocese of Arras. Soon after 1320 he obtained his master's degree at the University of Paris, where he was twice elected rector, first in 1328 and again in 1340.

The extant writings of Buridan consist of the lectures he gave at the University of Paris, where the curriculum was based largely on a study of Aristotle, along with textbooks on logic, grammar, mathematics, and astronomy. Buridan wrote his own textbook on logic as well as two advanced treatises on the subject. All of his other writings are commentaries and books of *Questions* on Aristotle's principal works.

Buridan's philosophy of science is enunciated in his *Questions* on Aristotle's *Physics* and *Metaphysics*. There he makes the distinction between premises whose necessity is determined through logic and those based on empirical evidence, whose necessity is conditional "on the supposition of the common cause of nature." He held that the principles of natural science are of the second type, noting that they "are not immediately evident; indeed we may be in doubt concerning them for a long time. But they are called principles because they are incommensurable, because they cannot be deduced from other premises nor be proved by any formal procedure; but they are accepted because they have been observed to be true in many instances and to be false in none."

This theory is significant for its rejection of the notion, held by most Latin commentators on Aristotle, that the principles of physics are necessary conclusions established by metaphysics. Buridan, by holding that scientific principles are inductive generalizations based on common experience, excluded supernatural causes as being irrelevant to the method and purpose of science. But at the same time he admitted the absolute possibility of divine intervention in the causal order of nature, and in this way he tried to avoid conflict with Christian dogma.

He took as a working hypothesis for natural philosophy the general framework of Aristotelian physics and cosmology, along with its interpretations by Greek, Islamic, and Christian commentators. By Buridan's time the authority of Aristotle had been challenged, not only because of conflict with Christian dogma, but also on the grounds that some of his theories were contrary to observed facts. Buridan's major significance in the history of science arises from a challenge that he made to Aristotle's explanation of projectile motion, in which he proposed an alternative theory that would prove to be one of the basic ideas of modern dynamics.

Buridan began by presenting and then refuting Aristotle's so-called antiperistasis theory for explaining why a projectile continues to move after it has lost contact with the source of motion. He then proposed his own explanation, the so-called impetus theory, the revival of a concept first proposed in the sixth century by John Philoponus. Buridan explains the continued motion of a projectile as being due to the impetus it received from the force of projection, and which "would endure forever if it were not diminished and corrupted by an opposing resistance or something tending to an opposed motion." He defines impetus as a function of the body's "quantity of matter" and its velocity, which is equivalent to the modern concept of momentum, or mass times velocity, where mass is the inertial property of matter, its resistance to a change in its state of motion. As applied to the case of free fall, Buridan explains that gravity is not only the primary cause of the motion, but also imparts additional increments of impetus to the body as it falls, thus accelerating it, that is, increasing its velocity:

> It must be imagined that a heavy body acquires from its primary mover, namely from its gravity, not merely motion, but also with that motion, a certain impetus such as is able to move that body along with the natural constant gravity. And because that impetus is acquired commensurately with motion, it follows that the faster the motion, the greater and stronger is the impetus. Thus the heavy body is moved initially only by its natural gravity, and hence slowly; but then it is moved by that same gravity as well as by the impetus already acquired, and thus it is … continuously accelerated to the end.

Truly revolutionary in its time, Buridan is essentially saying that a force, in this case gravity, gives the body successive increments of impetus. Since the mass of the body does not change and impetus is mass times velocity, this is equivalent to saying that force is that which changes the velocity of a body. Then if there is no force the velocity will not change but will remain constant. Thus Buridan's impetus theory is very close to two of Newton's laws of motion, in which the acceleration of a body is proportional to the force acting on it, and if the force is zero, so is the acceleration; that is, the velocity is constant, which is the principle of inertia. But it is not clear whether Buridan thought that the velocity of a falling body increased proportionately to the time elapsed, which is correct, or in proportion to the distance fallen, the latter being a mistaken view that was perpetuated into the seventeenth century.

Buridan's concept of impetus is further distinguished from the modern view in his mistaken view of rotational motion, where he believed that rotation at constant angular velocity was due to a rotational impetus analogous to the rectilinear impetus in projectile motion, an erroneous idea that was revived by Galileo.

Buridan proposed an interesting interpretation of elastic collisions using his impetus theory, one that is used in modern classical mechanics. Crombie gave a clear explanation of Buridan's model and its application to vibrations and oscillations:

> As to questions of terrestrial dynamics, Buridan himself explained the bouncing of a tennis ball by analogy with the reflection of light, by saying that the initial *impetus* compressed the ball by violence when it struck the ground, and when it sprang back this imparted a new *impetus* which caused the ball to bounce up. He gave a similar explanation for the vibration of plucked strings and the oscillation of a pendulum.

This concept of impetus would also explain the motion of the celestial spheres, which in Aristotelian cosmology rotated with constant velocity. Buridan argued that there was no need to have immaterial "intelligences" as the unmoved movers of the celestial spheres, as supposed by Aristotle, because their motion was inertial after the initial impetus they received from the Creator. "For it could be said that God, in creating the world, set each celestial orb in motion … and, in setting them in motion, he gave them an impetus capable of keeping them in motion without there being any need of his moving them any more." He added that this was why God could rest on the seventh day after the Creation, for the inertia of the celestial spheres would keep them in motion without the need for any additional divine effort.

Although Buridan differed from Aristotle in formulating his impetus theory and applying it to the motion of projectiles and freely falling bodies, he continued to use the Aristotelian theory of motion. Thus in commenting on Aristotle's argument against the existence of a void, Buridan accepted the notion that the velocity of natural motion in a medium is inversely proportional to the resistance of the medium, so that in a vacuum motion would be instantaneous, and since this is impossible the void does not exist.

Buridan makes an interesting point in his *Questions* on Aristotle's *Physics*, where he remarks on the inherent uncertainty of the description that mathematicians could give of the world of experience. "It is to be noted that we cannot measure natural motions absolutely precisely and punctually, that is according to the manner of mathematical considerations. We cannot know by means of a balance if a pound of wax is precisely equal to a pound of lead, for there could be an excess of so small a quantity that we might not be able to detect it."

In his *Questions* on Aristotle's *De caelo et mundo*, Buridan asks if a proof can be given for Aristotle's geocentric model, in which the earth is at rest at the center of the cosmos with the stars and other celestial bodies rotating around it. He noted that many in his time believed the contrary, that the earth is rotating on its axis and that the stellar sphere is at rest, adding that it is "indisputably true that if the facts were as this theory supposes, everything in the heavens would appear to us just as it now appears." In support of the earth's rotation, he says that is better to account for appearances by the simplest theory, and it is more reasonable to think that the vastly greater stellar sphere is at rest and the earth is moving, rather than the other way around. But, after refuting the usual arguments against the earth's rotation, Buridan says that he himself believes the contrary, using the argument that a projectile fired directly upward will fall back to its starting point, which is true, at least approximately, whether or not the earth is rotating.

Although he rejected the idea of the earth's rotation, Buridan differed from the Aristotelian idea that the earth is the immobile enter of the universe. Ernest A. Moody gives a summary of Buridan's interesting proof of this revolutionary theory:

> Because the dry land protruding from the ocean is mostly on one side of the earth, the center of volume of the earth does not coincide with its center of gravity. The earth, however, is in the center of the world in the sense that its center of gravity is equidistant from the inner surface of the celestial spheres. But this center of gravity is continuously altered by the erosion of the dry land which slowly gets washed into the sea; and consequently the whole mass of the earth slowly shifts from the wet side to the dry side in order to keep its center of gravity at the center of the heavens.

The impetus theory was adopted by Buridan's students and became known throughout Europe, though in a corrupted form that restored some Aristotelian notions. Aside from that, Buridan is credited with eliminating explanations involving Aristotelian final causes from physics. His books were required reading at universities until the seventeenth century, and would have been read by both Copernicus and Galileo. Copernicus used some of Buridan's arguments in discussing the earth's motions, and Galileo revived the theory of impetus in formulating his own laws of kinematics and dynamics.

Like other famous medieval scholars, Buridan was often the subject of apocryphal stories. One of these tales, perpetuated by the poet François Villon, tells of how Buridan had an affair with the wife of Charles V of France, who had him tied up in a sack and thrown into the Seine. Another story, known as that of "Buridan's ass," attributes to him the idea that an ass standing between two equally desirable bundles of hay would starve to death because he could not choose which one of them to eat.

One of the first scholars to be influenced by Buridan was Albert of Saxony (c. 1316–1390), who was teaching at the University of Paris at the same time. Albert was made rector of the university in 1353 and served until the end of 1362, when he went to Avignon to carry out several commissions for Pope Urban V. When the University of Vienna was established in June 1365, Albert became its first rector. A year later he was appointed bishop of Halberstadt, a post he held for the rest of his life. At one point he was accused of heresy by some of the clergy in his diocese, who charged that he was "more learned in human science than in divine wisdom."

Albert's lectures on natural philosophy at the University of Paris were patterned on those of Buridan. Albert's works thus served to transmit Buridan's ideas, particularly his explanations of projectile motion and of gravitational acceleration through the impetus theory, although he introduced an error that would lead to interesting discussions by his successors, as Ernest A. Moody explained:

> Unlike Buridan, he introduced an error into the analysis of projectile motion, by supposing that there is a brief period of rest between the ascent of a projectile hurled directly upward and its descent. Yet this led him to initiate a fruitful discussion by raising the question of the trajectory that would be followed by a projectile shot horizontally from a cannon. He supposed that it would follow a straight horizontal path until its impetus ceased to exceed the force of its gravity, but that it would then follow a curved path for a short period in which its lateral impetus would be compounded with a downward impetus caused by its gravity, after which it would fall straight down. Leonardo da Vinci took up the problem, but it remained for Niccolo Tartaglia to show that the entire trajectory would be a curve determined by a composition of the two forces.

The most distinguished of Buridan's students, though, was Nicole Oresme (c. 1320–1382), who studied under him at the University of Paris in the 1340s. Chosen grand master of the university's College of Navarre in 1356, he later became secretary of the dauphin of France, the future king Charles V. From about 1369 he was employed by Charles to translate some of Aristotle's Latin texts into French, for which he was rewarded in 1377 when, at the behest of the king, he was made bishop of Lisieux, a post he held until his death in 1382.

Miniature of Nicole Oresme

Oresme gave a geometrical proof of the Merton mean-speed rule for motion at constant acceleration in his *Tractus de configurationibus qualitatum et motuum*, written in the 1350s while he was at the College of Navarre in Paris. The proof is in the form of a graph in which the velocity (v) is on the vertical axis as a function of the time (t) on the horizontal axis. A body starting at rest and accelerating at a constant rate is plotted as a straight line sloping upward from the origin, forming a triangle, whereas one traveling at constant velocity would graph as a straight horizontal line, forming a rectangle.

Consider, for example, the case of a body accelerating so that its velocity increases from zero by two feet per second every second. Graphing the motion for four seconds, the velocity increases each second from 0–2–4–6–8 feet per second at the end of four seconds. This graphs in the form of a straight line rising from 0 to 8 feet per second over a time interval of four seconds, forming a right triangle with a height of 8 and a base of 4. The acceleration (a) is equal to the slope of the straight line, which is 8/4, or 2, in units of feet per second per second. The average velocity is half of the final velocity, which is 8/2, or 4 feet per second. The mean-speed rule then gives the distance s traveled in 4 seconds as 4 times 4, or 16 feet, which is equal to the distance covered in the same time by a body moving at a constant speed of 4 feet per second. The rule can be applied over each one-second interval, and one finds that the average velocity for each second increases from 1–3–5–7 feet per second. The distance in feet traveled in the first second is then 1, in the second 3, in the third 5, and in the fourth 7. These results can be generalized by the equations v = a times t, and s = a/2 times t squared. These are the kinematic equations formulated by Galileo in his *Dialogue Concerning the Two New Sciences* (1638), where he used Oresme's proof of the Merton mean-speed rule.

However, Oresme did not apply the mean-speed rule to the problem of a freely falling body, as Galileo was to do almost three hundred years later. Oresme did use Buridan's concept of impetus in explaining the acceleration of a falling body by successive increments of impetus due to the gravitational force. He seems to have concluded, correctly, that the velocity of a falling body is directly proportional to the time elapsed, counter to the prevailing opinion that it depended on the distance fallen. As a hypothetical example, he considers the case of a body that falls through a hole bored through a diameter of the earth, showing that when it reaches the center its impetus will carry it on until its acquired impetus is lost, whereupon it will fall back and oscillate up and down about the center, a problem that I discuss in my courses in elementary physics.

Oresme also had original ideas in astronomy, which he presented in his *Livre du ciel et du monde d'Aristote*, written in 1377 for Charles V. One of these was his comparison of the eternal motion of the celestial spheres to a perpetual mechanical clock, set in motion by God at the moment of their creation. He wrote that "it is not impossible that the heavens are moved by a power or corporeal quality in it, without violence and without work, because the resistance in the heavens does not incline them to any other movement nor to rest but only that they are not moved more quickly."

Oresme objected to the Aristotelian notion that the earth was the stationary center of the finite cosmos and the reference point for all motion and gravitation. He argued that motion, gravity, and the directions in space must be regarded as relative, saying that God, through his omnipotence, could create an infinite space and as many universes as he chose. Oresme was thus able to reject the idea that the earth was the fixed center of the cosmos. Instead he proposed the idea that gravity was simply the tendency of bodies to move toward the center of spherical mass distributions. Gravitational motion was relative only to a particular universe; there was no absolute direction of gravity applying to all of space. But although God had the power to make multiple worlds, Oresme ultimately rejected this possibility in favor of the earth-centered cosmos of Aristotle, as he says at the end of his discussion of gravitation:

> All heavy things of this world tend to be conjoined in one mass such
> that the center of gravity of this mass is in the center of this world,
> and the whole constitutes a single body in number. And
> accordingly they all have one [natural] place according to number.

And if a part of the [element] earth of another world was in this world, it would tend toward the center of this world and be conjoined with this mass.... But it does not accordingly follow that the parts of the [element] earth or heavy things of the other world (if it exists) tend to the center of this world, for in their world they would make a mass which would be a single body according to number, and which would have a single place according to number, and which would be ordered according to high or low [in respect to its own center] just as in the mass of heavy things in this world.... I conclude then that God can and would be able by his omnipotence to make another world other than this one, or several of them whether similar or dissimilar, and Aristotle offers no sufficient proof to the contrary. But as it was said before, in fact there never was, nor will there be, any but a single corporeal world.

Historian Marshall Clagett made an interesting observation about the above passage, noting "that it reveals the technique used by Oresme and his Parisian contemporaries, which permitted them to suggest the most unorthodox and radical philosophical ideas while disclaiming any commitment to them."

Oresme used the same technique in suggesting the relativity of motion and the diurnal rotation of the earth. He proposed, "subject to correction, that the earth is moving with daily motion and the heavens not. And first I will declare that it is impossible to show the contrary by any observation; secondly from reason: and thirdly I will give reasons in favor of the opinion."

Oresme's arguments in favor of the earth's motion would later be used by both Copernicus and Galileo. Despite all of these arguments, Oresme, who at the time had just been appointed bishop of Lisieux, in the end rejected the idea of the diurnal rotation of the earth as being contrary to Christian doctrine, "for God fixed the earth, so that it does not move, notwithstanding the reasons to the contrary." Orseme's attitude was not uncommon among his clerical contemporaries, for in his position as bishop he was sworn to uphold the doctrines of the Catholic Church, even when they conflicted with his own philosophical ideas.

The third part of the *Tractus de configurationibus* is particularly interesting because of Oresme's distinction between convergent and diverging series. Here he states that when an infinite series is such that to a given magnitude there are added "proportional parts to infinity" in which the ratio of the parts is less than one, the series has a finite sum, otherwise "the total would be infinite." This and other mathematical work of Oresme's have earned him a place as one of the pioneers in the development of calculus, three centuries before Newton and Leibniz.

Oresme's insistence on rational explanation of natural phenomena rather than supernatural is particularly evident in his *Questio contra divinatories*. There he says that because of ignorance men attribute apparently fabulous events to the heavens, to God, or to demons, which leads to the "destruction of philosophy." Concerning the existence of demons, he goes on to say, "Moreover, if the Faith did not pose their existence, I would say that from no natural effect can they be proved to exist, for all things [supposedly arising from them] can be saved [i.e., explained] naturally."

Oresme was the last distinguished thinker of the medieval Oxford-Paris school, which went into decline toward the end of the fourteenth century, when, according to Crombie, "original writing in Oxford and Paris seems practically to have ceased for nearly two hundred years."

It took more than a century after the decline of the Oxford-Paris school for another revolutionary scientist to emerge in western Europe: Leonardo da Vinci (1452–1519). Unfortunately, his extraordinary artistic accomplishments have overshadowed his achievements in science, in which he made important contributions in anatomy, geology, hydrostatics, optics, technology, mechanics, and dynamics. Leonardo's dynamics was based on the theory of impetus, and he accepted Albert of Saxony's explanation of projectile motion, though he recognized that the actual motion of a missile might be the resultant of two or more different forces, such as that of gravity and of the propellant. He understood that the velocity of a ball rolling down an inclined plane was uniformly accelerated, that is, that its speed increases by equal amounts in equal intervals of time. He also showed that the velocity of a ball falling vertically from a certain height would be the same as if rolled down an inclined plane from the same height, though it would take a longer time. Both of these are discoveries that are credited to Galileo. Leonardo also gave the correct law for the acceleration of a freely falling body, although, as Crombie put it, he did so "with considerable confusion."

The foundations of the work of Leonardo and subsequent writers on dynamics up through Galileo had been laid by the scholars of the Oxford school in the thirteenth and fourteenth centuries. They applied their ideas on methodology to the study, taking a radically new approach to the study of a body moving in space and time. This enabled them to write the first laws of kinematics, giving a purely mathematical description of motion, anticipating the work of Galileo. They also took a new and dynamic approach to states of equilibrium, such as the law of the lever.

Beginning in the fourteenth century, scholars in Paris, most notably Jean Buridan and Nicole Oresme, used the new methodology to study dynamics, the relation between force and motion. The most important idea in the new dynamics was that of impetus, first proposed by John Philoponus, and now developed to explain the trajectories of projectiles, the accelerated motion of falling bodies, the motion of pendulums and bouncing balls, and the orbital motions of the planets. Also of great importance were the concepts of gravity as a force and of the persistence of impetus, the latter leading directly to that of the inertia of matter, Newton's first law of motion.

The study of accelerated motion also began in the fourteenth century, and the kinematic equations for a body moving with constant acceleration was applied to the motion of falling bodies, where gravity was identified as the force that provided the added impetus, which would eventually lead to Newton's second law of motion.

The discussion of the new kinematics and dynamics cast up a number of new and revolutionary ideas: that the universe might be infinite, that there might be many universes, and the relativity of motion, so that the earth might be rotating and the heavens stationary.

Reviewing all of this, I realize that the advances made by the Oxford-Paris school in the science of motion during the thirteenth, fourteenth, and fifteenth centuries still make up a large part of the university courses that I have taught in elementary physics, which always begins with kinematics and dynamics, the science of motion, for everything else depends on that.

9

Over the Rainbow

*N*ATURE ITSELF WAS THE FIRST PHYSICS LABORATORY, AND AMONG natural phenomena there are none that are more beautiful and spectacular than the rainbow. It cries out for an explanation, and long before science this took the form of myths, with the ancient Greeks imagining that the rainbow was the path made by the goddess Iris when Zeus sent her to earth with a message. Being Irish, I was told by my paternal grandfather Tomas that the leprechaun's secret hiding place for his pot of gold is at the end of the rainbow, but he didn't tell me which end.

Aristotle was the first to attempt a scientific explanation of the rainbow, and his work formed the basis for subsequent research in medieval Islam and Latin Europe. Robert Grosseteste's work on the rainbow, though it was mistaken in some of its basic ideas, was the starting point for those who followed him in their study of optics, some of their experiments being designed to simulate the rainbow.

We usually see a rainbow in the morning or evening after a rain shower; standing with our back to the sun, it appears as an arc with the full spectrum of visible light ranging from red on the outside to violet on the inside. The center of the bow is exactly opposite the sun relative to the observer, and if you move forward or backward the rainbow moves with you, so the leprechaun's pot of gold remains safe, or so Tomas told me. If you have a friend with you, he has his own rainbow that moves with him. The arc makes an angle of 40 to 42° with the direction opposite the sun.

This is the primary rainbow. There is often a fainter secondary rainbow visible outside the primary arc, making an angle of 50 to 53°. The spectrum of colors in the secondary rainbow is the reverse of that in the primary arc, with violet on the outside and red on the inside. The dark region between the primary and secondary rainbows is known as Alexander's band.

As we shall see, during the thirteenth and early fourteenth centuries a succession of scholars beginning with Grosseteste did research on the rainbow, laying the foundations for the theory that Newton published early in the eighteenth century. There, as we learned, Newton explained the rainbow as being due to a combination of reflection, refraction, and dispersion, or the division of sunlight into its component colors, a continuous spectrum extending from red to violet. The rainbow is due to sunlight passing through individual droplets of rain remaining from a shower. In the primary rainbow the light is refracted when it enters the drop, reflected internally at its back surface, and then refracted again as it leaves. In the secondary rainbow the light undergoes two internal reflections, inverting the spectrum of colors.

Grosseteste's work on the rainbow inspired some verses written about 1270 by the French poet Jean de Meun in his continuation of Guillaume de Lorris's *Romance of the Rose*. These are in chapter 83, where *Nature explains the influence of the heavens*, telling of how the clouds, "to give solace to the earth":

> Are wont to bear, ready at hand, a bow
> Or two or even three if they prefer,
> The which celestial arcs are rainbows called,
> Regarding which nobody can explain,
> Unless he teaches optics at some school,
> How they are varicolored by the sun,
>
> How many and what sorts of hues they show,
> Wherefore so many and such different kinds,
> Or why they are displayed in such a form.

The first significant advance beyond what Grosseteste and Bacon had done was by the Polish scholar Witelo (c. 1230–c. 1275). What little is known of Witelo's life comes from scattered references in his best-known work, the *Perspectiva*. There he refers to "our homeland, namely Poland," and he mentions Wroclaw (Breslau) and towns in its vicinity, indicating that he was born and raised there. He seems to have done his undergraduate studies in Paris, as surmised from his description of a nocturnal student brawl there in 1253. These battles were frequent occurrences at medieval universities and often involved the townspeople as well, as in the town versus gown brawl at Oxford in 1209 that closed the university for five years.

A decade later he is referred to as "Witelo, student of canon law" in a treatise he wrote while doing graduate studies at the University of Padua. During the winter of 1268–1269 Witelo appeared in Viterbo, where he met William of Moerbeke, the famous translator of scientific works from Greek to Latin, to whom he dedicated his *Perspectiva*. A manuscript of the *Perspectiva* refers to the author as "Magister Witelo de Viconia," which has led to the suggestion that he retired to the abbey of Vicogne in his later years.

The *Perspectiva* is based on the optical works of Robert Grosseteste and Roger Bacon as well as those of ibn al-Haytham, Ptolemy, and Hero of Alexandria. It would seem that the *Perspectiva* was not written before 1270, since it makes use of Hero's *Catoptrica*, the translation of which was completed by William of Moerbeke on December 31, 1269. The *Perspectiva* is such an immense work, coming to more than five hundred pages in the three printed editions, that it probably took several years to write, and so it would seem as if Witelo lived on until at least the mid-1270s.

Witelo adopted the ideas of the "metaphysics of light" and the "multiplication of forms" directly from Grosseteste and Roger Bacon. He said in the preface to the *Perspectiva*, "Sensible light is the intermediary of corporeal influences"; "light is the corporeal form"; and "Light is the first of all sensible forms." The latter remark implies that light is an intermediary in certain phenomena, a case of the multiplication of forms. David C. Lindberg remarked that "although light is only one instance of natural action, it is the instance accessible to the senses and most amenable to analysis; therefore it serves, for Witelo, as the paradigm for the investigation of all natural actions."

Thus Witelo, in discussing refraction and the dispersion of light into a spectrum of colors in the *Perspectiva*, wrote, "These are the things that occur to lights and colors in their diffusion through transparent bodies and in the refraction that occurs in all of them." And in the preface, discussing in general the action of one body on another, he noted that "the investigation properly proceeds by means of visible entities."

In the preface to the *Perspectiva*, which he addressed to William of Moerbeke, Witelo wrote "of corporeal influences sensible light is the medium," adding that "there is something wonderful in the way in which the influence of divine power flows in to things of the lower world passing through the powers of the higher world."

Witelo disagreed with Grosseteste and Bacon where they say that light rays travel from the observer's eye to the visible object, and instead followed the reverse theory of ibn al-Haytham's *Optics* that the rays emanate from the luminous object to interact with the eye, which we now know to be correct.

The *Perspectiva* is in ten books, the first of which consists of definitions, postulates, and geometrical theorems, from which Witelo derived the mathematical principles needed for his optical demonstrations in the following nine books. These cover the whole of what we now call geometrical optics, including reflection by plane and curved mirrors and refraction at plane or spherical interfaces between two different media, along with the rainbow and other phenomena occurring in the atmosphere.

It also covers the physiology, psychology, and geometrical optics of vision, both monocular and binocular; visual perception of size, shape, distance, roughness, darkness, and even visual beauty; along with the nature of radiation, the propagation of light, white and colored, the formation of shadows, and the camera obscura, or pinhole camera. The latter, invented by ibn al-Haytham and introduced to Europe by Witelo, is the basis of the modern camera, one of the most important inventions of medieval science.

Witelo followed Hero of Alexandria in deriving the law of reflection from the principle that a ray of light, in being reflected, will follow the path of minimum length, saying that nature does nothing in vain and "always acts along the shortest lines."

Geometric optics is based on the principle that light is propagated rectilinearly, changing direction only when reflected or refracted at an angle of incidence other than zero. Witelo drew a distinction between the one-dimensional Euclidian lines used in a geometrical demonstration of optical analysis and actual light rays, which have perceptible width. He wrote: "In the least light ray that can be supposed, there is some width." Nevertheless, "in the middle of that [radial line] is an imaginary mathematical line, parallel to which are all the other mathematical lines in that mathematical line." This is the first indication of the wavelike nature of light, which causes it to spread out as it propagates through space, an idea that was revived by Christiaan Huygens (1629–1695) and Newton. It led to a discussion about whether light was a wave or a particle, which was finally resolved by Louis de Broglie (1892–1987), who won the Nobel Prize in Physics in 1929 for his wave-particle duality theory, which states that any moving particle has an associated wave, the basis of modern wave mechanics.

The methodology of Witelo is distinguished by his use of experimentation with quantitative results subjected to mathematical analysis. A good example is his attempt to construct a burning mirror that would concentrate the sun's rays at a single focal point. After discussing various possibilities, he describes a paraboloidal mirror that he claimed to have invented himself. This was a surface produced by the rotation of a parabola about its axis of symmetry, based on the properties of parabolas in the *Conics* of Apollonius. Crombie gave a summary of Witelo's description of how he manufactured a paraboloidal mirror from a concave piece of metal:

> Two equal parabolic sections were drawn on a rectangular sheet of good iron or steel, and cut out. The parabolic edge of one was sharpened for cutting and that of the other made like a file for polishing. These sections were then used, with some mechanism to rotate them about their axes, to cut and polish the concave surface of the piece of iron so that it formed a parabolic mirror.

Another important example of Witelo's method is his work on refraction. Here he made most of his measurements by repeating, with some changes, experiments described by ibn al-Haytham, whose work was based on similar experiments by Ptolemy. Ibn al-Haytham corrected Ptolemy by showing that the angle of refraction was not simply proportional to the angle of incidence. He also appears to have been the first to state the "reciprocal law," which says that a light ray crossing the interface between two media will, if reversed, follow the same path. Witelo, in making a statement about the refraction of light, recommends that anyone wishing to study this phenomenon should use the instrument described by ibn al-Haytham, noting that "proof of this proposition depends on experiments with instruments rather than on other types of demonstration."

The instrument that Witelo describes comprised a cylindrical brass vessel with a metal rod passing through the center of its base and fixed to the sidewalls of a glass container parallel to its bottom. The wall of the cylindrical vessel was perforated by two holes at either end of the diameter of a circle marked off into 360° and further divided into minutes of arc. When used to study the refraction of light passing from air to water, the glass container was filled with water up to the rod that formed the central axis of the brass cylinder, which was rotated so that the diameter joining the two holes was perpendicular to the surface of the water, the upper hole acting as a sight. The rod was then rotated so that the sight made an angle with the perpendicular ranging from 10 to 80°, at intervals of 10°. Light from the sun or a candle was allowed to pass through the sight along the diameter of the graduated circle until it reached the surface of the water, where it was bent away from its original direction through refraction. The axis of the cylinder was then rotated until the lower hole lined up with the refracted ray, which could then be seen as a point of light. The angles of incidence and refraction, both measured with respect to the perpendicular, could then be measured off from the graduated circle.

Then, to measure the refraction of light passing from water to air, Witelo looked through the sight and moved a stylus around the graduated circle until it came into view. Thus he was observing the angle of incidence necessary for the rays coming from the stylus to pass through the sight when the latter was in a given position.

To study the refraction of light passing from air into glass, Witelo inserted a glass hemisphere into the lower half of the brass cylinder. Then, to measure the refraction of light passing from glass into air, he simply rotated the cylinder 180° so that the glass hemisphere was in the upper half rather than the lower.

He tabulated his data to show the angles of refraction for given angles of incidence for four different arrangements: light passing from air to water, water to air, air to glass, and glass to air. An analysis of his results for light passing between air and water has been made to compare them with the values found using the modern law of refraction. It is impossible to do the same for the other cases since Witelo did not identify the kind of glass he used, the index of refraction differing appreciably from one type to another.

For the case of light passing from air into water, the analysis shows reasonable agreement between the values for the angles of refraction found by Witelo and those obtained using the modern law of refraction. But this is not true for the case of light passing from water into air, where it seems as if Witelo did not actually make the observations he records but simply obtained the values by using the reciprocal law, which he did incorrectly. Also, the results he records for angles of incidence ranging from 50 to 80° are impossible, because above a critical angle no refraction takes place when light is passing from a denser medium to one that is less dense optically, a phenomenon known as total internal reflection, of which Witelo was apparently unaware.

Witelo tried to express his results in a number of mathematical generalizations, but his efforts represented little improvement on what had been done by his predecessors. He also used his instrument to show that light rays of different color travel in the same straight line as white light when passing through a single uniform medium. He did this by putting pieces of colored material in the path of rays of sunlight entering the sight, observing that they passed through the hole on the opposite side with deviation.

He went on to discuss the properties of convex and concave lenses, of which he appears to have had only a theoretical knowledge, his principal source being ibn al-Haytham. His analysis of refraction makes use of the concept later known as the principle of minimum path. He justified this by the metaphysical notion of economy, saying that "it would be futile for anything to take place by longer lines, when it could better and more certainly take place by shorter lines." In his analysis Witelo resolves the oblique motion of light into components perpendicular and parallel to the refracting surface, a technique that had come into use in the study of both optics and the science of motion.

In applying his knowledge of refraction to the study of the rainbow, Witelo drew heavily on the work of Roger Bacon. He differed from Bacon mainly in his belief that both reflection and refraction of sunlight in the individual raindrops of a cloud were involved in the phenomenon. Two observations convinced him of this. The first was that the rays by which the rainbow is seen came to the eye at angles equal to those at which the incident rays struck the drops, showing that reflection was involved. Secondly, unlike a simply reflected image, when an observer approached or retreated from the rainbow, it retreated or followed after him, respectively, without changing in size, indicating that refraction was also involved. He concluded, "Therefore the rainbow is seen not only by reflection but by the refraction of light within the body from which it is reflected." Crombie described Witelo's theory of the rainbow and the demonstration that he gave to support it.

It was seen, in fact, as part of the circumference of a cone produced by the rays coming to the eye, which was on the axis of the cone. As the observer approached or retreated from the rainbow, or moved to the right or left, he placed his eye on the axis of a different cone, of which an infinite number were produced by the multitudinous rays of light reflected from water drops in the atmosphere. This theory he supported by a simple experiment in which a rainbow in an artificial spray was seen to change its position according to whether the observer closed one eye or the other.

Witelo formulated a model to show how the water drops in the cloud produced the rainbow through a combination of refraction by individual raindrops and then reflection by other drops. His model would not, in fact, produce the results that he expected, but it paved the way for subsequent attempts to formulate the correct theory of the rainbow.

His study of the rainbow led Witelo to perform a number of experiments involving the refraction of sunlight by crystals. He produced the colors of the spectrum by passing light through a hexagonal crystal, observing that the blue rays were refracted more than the red. He also produced an artificial rainbow by passing sunlight through a spherical flask filled with water, noting that the subject was unexplored and experiment was the guide: "For the color or visible form is carried to vision only by the nature of light which it contains; and to what has been said the careful inquirer will be able by experiment to add many things."

According to Lindberg, Witelo described the eye "as a composition of three glacial humors—glacial or crystalline, vitreous, and albugineous (aqueous) and four tunics—uvea, cornea or retina." Witelo's description of these humors and tunics was predominately geometrical; all of them are spherical in form, and those behind the glacial humor are concentric so that a perpendicularly incident ray will pass through all of them without refraction. Furthermore, the glacial and vitreous humors have exactly the shapes and relative densities so that the rays will converge at the center of the eye. Sight occurs when there is a "union of the visible forms and the soul's organ" on the surface of the vitreous humor, or crystalline lens. The forms then pass through the optic nerve to the anterior part of the brain, where the nerves from the two eyes intersect to form the "common nerve," the site of the *ultimum sentiens*, or ultimate perception. Lindberg gives an interesting assessment of the immense influence of Witelo, whose works were linked with those of ibn al-Haytham and Witelo's English contemporary John Pecham:

It is difficult to separate Witelo's influence on the history of late medieval and early modern optics from that of ibn al-Haytham, particularly after their works were published in a single volume in 1572. One can affirm in general that their writings, along with John Pecham's *Perspectiva communis*, served as the standard textbooks on optics until well into the seventeenth century. More specifically, it is possible to establish Witelo's influence on Henry of Hesse, Blasius of Parma and Nicole Oresme in the fourteenth century; Lorenzo Ghiberti, Johannes Regiomontanus, and Leonardo da Vinci in the fifteenth century; Giambattista della Porta, Francesco Maurolyco, Giovanni Battista Benedetti, Tycho Brahe, William Gilbert, Simon Stevin, and Thomas Harriot in the sixteenth century; and Kepler, Galileo, Willebrord Snell, Descartes, and Francesco Grimaldi in the seventeenth century.

John Pecham (c. 1230–1292) was born in the vicinity of Lewes in Sussex, where he received his elementary education at the local priory. He later matriculated in the arts faculties at the universities of Oxford and Paris. He joined the Franciscan order in the 1250s and was sent to Paris to study theology, receiving a doctorate in 1269. During the years 1269–1271 he served as regent master in theology in Paris, after which he returned to Oxford to serve as lecturer in theology to the Franciscan school, a position he held until he was appointed provincial minister of the order in 1275. Two years later he went to Rome as master in theology to the papal curia; then in 1279 he was appointed archbishop of Canterbury, a post he held until his death on December 8, 1292.

Both Robert Grosseteste and Roger Bacon directly influenced Pecham's work in optics. Pecham was personally acquainted with Bacon, and they resided together in the Franciscan priory in Paris during the period when Bacon was writing his principal scientific works. Pecham's was also influenced by Aristotle, Euclid, Ptolemy, Augustine, al-Kindi, and ibn al-Haytham, whose *Optics* was the primary source for his work on the science of light.

Pecham's first optical work was the *Tractus de perspectiva*, probably written for the Franciscan schools during his years as a teacher in Paris and Oxford, since he says in the introduction that he wrote it to discuss light and number "for the sake of my simpler brothers." It has been dismissively described as "a rambling piece of continuous prose ... filled with quotations from the Bible and patristic sources, especially Augustine, that give it a theological and devotional flavor."

Perhaps Pecham's most important optical work is the *Perspectiva communis*, probably written in the years 1277–1279. He wrote in the introduction that his objective in writing this work was to "compress into concise summaries the teachings of perspective, which [in existing treatises] are presented with great obscurity." The *Perspectiva* is a clear and concise summary of the science of light at the time, based largely on ibn al-Haytham, Witelo, and Pseudo-Euclid's *De speculis*. It was deeply influenced by Grosseteste and Roger Bacon, in that Pecham conceived of light as a form of "multiplication of species," and regarded the study of optics as a way of introducing mathematical certainty into physics. There is nothing particularly original in the book, but it remained a popular text on optics until the seventeenth century, published in twelve printed editions between 1482 and 1665, used and cited by many medieval and Renaissance scholars, including Leonardo da Vinci and Kepler.

The *Perspectiva* begins with a description of several experiments illustrating the properties of light, as well as discussion topics such as the cause of reflection, the anatomy of the human eye, and the physiology of vision. It goes on to give a summary of the theory of concave mirrors, including a sophisticated analysis of image formation, followed by a summary of the theory of concave and convex lenses, concluding with a summary of current theories of the rainbow. Pecham's discussion of the rainbow attempts to reconcile previous theories of the phenomenon, where he argues that the effect is due to the concurrence of rectilinear, reflected, and refracted rays.

Pecham's other works on natural philosophy and mathematical science include *Tractatus de sphera* and *Tractatus de numeris*. The *Tractatus de sphera* gives an elementary discussion of cosmology, astronomy, and astrology, including such topics as the sphericity of the earth and the celestial bodies, the rotation of the heavens, the variation in the length of days, the climactic zones and the terrestrial sphere, and the causes of eclipses. The *Tractatus de numeris* deals with the elementary properties of number as well as their mystical significance, including an explanation of the Trinity.

The next advances in optics were made by Dietrich of Freiburg (c. 1250–c. 1311), who is sometimes called Theodoric. Dietrich, who is thought to be from Freiburg in Saxony, entered the Dominican order and probably taught in Germany before studying at the University of Paris in around 1275–1277. In 1304 he attended the general chapter of the Dominican order in Toulouse, where the master general, Aymeric de Plaisance, asked him to write up his researches on the rainbow. The result was his principal work, *De iride et radialibus impressionibus*, or *On the Rainbow and Radiant Impressions*, the latter term meaning phenomena produced in the upper atmosphere by radiation from the sun or any other celestial body.

The only Latin source cited by Dietrich is Albertus Magnus, but his treatise indicates that he was certainly aware of the work of ibn al-Haytham and Witelo, probably that of Roger Bacon, and perhaps that of Grosseteste. The Greek and Arabic sources that he cites are Aristotle's *Meteorologica* and other works, Euclid's *Elements*, the *Sphaera* of Theodosius, Ptolemy's *Almagest*, and works by Avicenna and Averroës. He was deeply influenced by Grosseteste and his followers, particularly in his belief in the "metaphysics of light" and in his scientific method, combining experiment and geometry with the principles of falsification and economy. He begins his lesser work, *De iride*, with the statement that knowledge of the rainbow, a phenomenon of admirable beauty, was guaranteed "by the combination of various infallible experiments with the efficacy of reasoning, as is clear enough from what follows."

De iride begins with an account of what Dietrich terms the three modes of visual apprehension, that is, through direct, reflected, and refracted rays. He then lists fifteen types of *impressiones radiales* with which he was familiar, including primary rainbows, secondary rainbows, white and colored halos around the sun, and colors in stars seen through a mist. He goes on to identify five types of optical phenomena involved in these effects: a single reflection, a single refraction, two refractions and a single internal reflection in a drop of water or a crystalline sphere, two refractions and two internal reflections in the drop or sphere, and total reflection at the boundary between two transparent media. He is the first to mention the phenomenon of total internal reflection, which occurs above a critical angle of incidence when light strikes the interface between two media, in the case where the optical density of the second medium is less than that of the first, such as from water to air. This is extremely important, for it is the principle that led to modern fiber optics, among other advances.

The remainder of *De iride* is devoted to the application of Dietrich's theory to the primary rainbow, the secondary rainbow, and the halo and other types of *impressiones radiales*, respectively.

Dietrich was the first to realize that the rainbow is due to the individual drops of rain rather than the cloud as a whole. This led him to make observations with a glass bowl filled with water, which he used as a model raindrop, for he writes "that a globe of water can be thought of, not as a diminutive spherical cloud, but as a magnified raindrop." He also did similar experiments with hexagonal crystals and spherical crystalline balls. His observations and geometrical analysis led him to conclude that light is refracted when it enters and leaves each raindrop, and that it is internally reflected once in creating the primary bow and twice for the secondary arc. In the primary rainbow the light enters the lower part of the drop and emerges at the upper, whereas in the secondary rainbow it is just the opposite. As a result the orders of the colors in the spectrum are reversed in the two bows, ranging from red in the outer arc to blue in the lower for the primary or lower rainbow, and from blue in the outer arc to red in the lower for the secondary or outer bow.

Dietrich drew a series of geometrical diagrams to show "the manner in which the colours which appear in the rainbow come to the eye, in the case of the lower rainbow." He summarized the results of his experiments to show that the collection of drops in which the rainbow appeared had "the breadth of the whole rainbow and also of each of its separate colours," and "that the place of incidence of the radiation into any drop" and the places of internal reflection and of emergence had "breadth according to the different parts of which breadth the different colors shine so that any one of such parts has a breadth corresponding to one of the colors." Furthermore, "in proportion as the radiation is incident along an oblique line more removed from the perpendicular, so does it go forth to the eye along another oblique line more removed from the same perpendicular."

He went on to explain the manner in which the component colors come to the eye from drops in different positions:

> So all the colors do not come spontaneously to the eye when it is in one and the same position with respect to the drop, but different colors come to the eye according to the different positions in which it is put with respect to a particular drop. And so if all the colors are seen simultaneously, as happens in the rainbow, this must necessarily result from different drops which have different positions with respect to the eye and the eye to them.

Referring to the first of a series of geometrical drawings, he shows how the banded arcs of the different colors appear as they do in the primary or lower rainbow, in which light enters the lower side of the drops and, after a single internal reflection, leaves through the upper. Then in his summary he gives the order of the colors in the rainbow, beginning with the outermost arc.

> And so all the colors of the rainbow are seen at the same time and the whole rainbow appears in the circle of altitude in different little spherical drops according to which they are more or less elevated to different parts of the arch, from which particular parts particular colors come to the eye in the manner described. But from the drops elevated in the circle of altitude above the said arc no incident radiation is sent to the eye. The drops depressed below the said arc

send some radiation incident upon them to the eye, but not with the colors of the rainbow but with white light unmixed with color.... Therefore, for the reasons stated, the color red shines in the highest part of the circle of altitude, next to this yellow, thirdly green, and finally blue, it follows that the upper and outer circle is blue, the next below yellow, then follows green, and the lowest and inner circle is blue.

Dietrich referred to another drawing to explain the secondary rainbow, in which the light enters on the lower side of the drop and, after making two internal reflections, emerges on the upper side. He showed that the arc of the secondary rainbow had the same center as that of the primary bow, with the entire secondary rainbow 11° higher than the lower bow.

Dietrich erred in his assertion that when the sun was on the horizon the maximum altitude of the primary rainbow was only 22°, as compared to the correct value of 42°, as Roger Bacon and Witelo had written. Dietrich made a number of other errors in his analysis; nevertheless, his work on the rainbow was far superior to those of any of his predecessors, and it paved the way for researches by his successors, most notably Regiomontanus, Descartes, and Newton.

Dietrich's theory of the rainbow is very similar to that of his Persian contemporary, Kamal al-Din al-Farisi. Dietrich does not cite the work of al-Farisi, but since it was never translated from Arabic into Latin, he was probably not aware of it. In any event, it seems that the emerging European science had by the beginning of the fourteenth century reached a level comparable to that of Arabic scientific research, at least in optics. But whereas the work of al-Farisi was the last great achievement of Arabic optics, Dietrich's researches would be an important stage in the further development of European studies in the science of light, culminating in the first correct theories of the rainbow and other optical phenomena in the seventeenth century by Newton, the final solution of a problem that had almost been solved fourteen centuries earlier by Claudius Ptolemaeus of Alexandria.

The very complexity of the rainbow, its geometry as well as its physics, made it an ideal subject for medieval physics to cut its teeth on, and thus optics, the science of light, developed apace with dynamics, the science of motion. As Crombie pointed out, "The work undertaken to explain the rainbow became linked with other work on originally independent problems." And as we have seen, research on the rainbow in the thirteenth and early fourteenth centuries involved a number of innovations: the use of magnifying lenses and the invention of spectacles; an attempt to derive the laws of reflection and refraction from the principle of minimum path; the use of conic sections in the theory of paraboloidal mirrors; Witelo's application of his studies in refraction to the optics of the human eye, and his realization that light rays have a finite width, a discovery that would lead to modern wave mechanics; Dietrich's researches on the dispersal of sunlight in water-filled glass spheres into its spectrum of colors, anticipating Newton's discoveries with glass prisms, and his discovery of total internal reflection, with its modern application in fiber optics. As Crombie concluded from all this activity: "To the eventually accepted theory of the rainbow nearly all this work contributed. Each separate contributor saw as the goal of contemporary optics as a whole the building up of a general theory of light and color by means of which all the separate problems would be related in a single system."

When Newton's successful explanation of the rainbow finally appeared in the first edition of his *Opticks* in 1714, his theory was a direct result of the researches done in the thirteenth and fourteenth centuries. A little more than a century later John Keats reacted to Newton's theory in these lines from his "Lamia":

> Do not all charms fly
> At the mere touch of cold philosophy?
> There was an awful rainbow once in heaven:
> We knew her woof, her texture: she is given
> In the dull catalogue of common things.
> Philosophy will clip an Angel's wings,

Conquer all mysteries by rule and line.
Empty the haunted air, and gnomed mine—
Unweave a rainbow.

One of the ironies of modern physics, I feel, is that the theory of the rainbow, one of the triumphs of medieval physics and a big step forward on the road to modern science, is no longer taught in elementary physics courses, and so most people today have no better understanding of it than my grandfather Tomas. And whereas Tomas never missed a rainbow, for he would be out fishing when they appeared, almost everyone I know today hardly ever sees one, and so they will never find the leprechaun's pot of gold.

10

The Revival of Astronomy, East and West

*T*HE DEVELOPMENT OF EUROPEAN ASTRONOMY ACCELERATED WITH the acquisition of Greek and Arabic texts on the subject translated into Latin, which were used in the universities that were founded beginning in the second half of the twelfth century. The masters in the universities prepared new texts for their students in their various disciplines, including astronomy, for which there had been little available previously. Astronomy proved to be particularly popular, and by the middle of the thirteenth century there were so many texts in the subject that Oliver of Brittany declared that "a day would scarcely suffice to completely tell of [astronomy's] innumerable books and authors."

A study of the surviving texts clearly shows that many masters went beyond the minimum requirements of the curriculum to give a full introduction to astronomy. According to historian Stephen C. McCluskey: "They collected these texts into a single unit, the *corpus astronomicum*, which covered computus, the calendar, the nature of the celestial spheres, planetary theory, and the use of astronomical instruments. Some works in the *corpus* were selected from the new translations, but most were works written in the beginning of the [thirteenth] century by western European authors."

These new astronomical texts also included calendars containing much of the same material as in earlier liturgical calendars, such as saints' days with tables of the motion of the sun through the zodiac, the length of daylight, and phases of the moon, but using new theories and observations to give more accurate values for these astronomical data. The texts usually included the so-called *algorismus*, which explained the arithmetical techniques useful in astronomical calculations, as well as explaining timekeeping by the sun and stars along with the use of the astrolabe and astronomical quadrant.

The most important of these new astronomical texts went by the general title of *De sphaera*. One of the earliest of these was Robert Grosseteste's *De sphaera*, published in 1215, in which, as we have seen, he discussed elements of both Aristotelian and Ptolemaic theoretical astronomy. Here Grosseteste gave an introduction comprising a geometrical description of the structure and motions of what he and his contemporaries called the *machina mundi*, the machine of the world. He discussed topics such as the varying length of the day in far more detail than earlier Latin writers, including not only the sphericity of the earth in his explanation but also the complexities due to the eccentricity of the ecliptic and the variation of the sun's motion caused by the inclination of the zodiac, which he gave in the newly introduced standard degrees and minutes as 24°33'. His discussion of lunar eclipses includes a clear explanation of the effect of the moon's inclined path, stating that for an eclipse to occur the moon must be within 12° of one of the lunar nodes, the points where the lunar and solar orbits intersect. He explained phenomena such as the nightly rotation of the stars around the fixed pole star, and showed why in traveling north the observer sees the northern constellations higher in the sky. Sometimes he went beyond geometrical demonstration to give physical explanations, as in presenting Aristotle's theory that since the earth is a heavy body its parts naturally gravitate toward the center to form a sphere.

Grosseteste's *De sphaera* was ideally suited for students who wanted to learn the broad geometrical basis for the motion and visibility of the sun, moon, and stars, but not for the complicated motions of the five planets. He gave the theoretical basis for astronomical timekeeping and computus, but in doing so he rejected Ptolemy's theory of the precession of the equinoxes (discovered by Hipparchus), in favor of the erroneous trepidation theory from Arabic astronomy.

As a result, Grosseteste's *De sphaera* was very demanding, and it opened the way for a competing text of the same title written at about the same time by his contemporary John of Holywood, better known by his Latin name, Johannes de Sacrobosco.

The little that is known of Sacrobosco's life is to a large extent speculation. He was probably born toward the end of the twelfth century and became a monk at the Augustine monastery of Holywood in Nithsdal, in Yorkshire. He is believed to have studied at Oxford before leaving around 1220 for Paris, where on June 5, 1221, he was admitted to the university under the syndics of the Scottish nation. Soon afterward he was elected professor of mathematics and subsequently won renown as an astronomer and mathematician, being among the first to teach and write about the new Arabic arithmetic and algebra. He died in either 1244 or 1256 and was buried in the convent of St. Mathurin in Paris. The uncertainty in his date of death stems from the ambiguity of the epitaph on his tombstone, which is decorated with the carving of an astrolabe, commemorating him as someone who had devoted his life to astronomy and mathematics.

Sacrobosco's principal extant works are three elementary textbooks on astronomy and mathematics: *De sphaera*, *De computo eccliastico*, and *De algorismo*, all of which are frequently bound together in the same manuscript. Sacrobosco's fame is principally based on his *De sphaera*, an astronomy text based on Ptolemy and his Arabic commentators, most notably al-Farghani. The text was first used at the University of Paris and then at all schools throughout Europe, its appeal being that it was simpler and clearer than Grosseteste's *De sphaera* and needed little or no explanation. As McCluskey pointed out:

> Like Grosseteste, Sacrobosco discussed the terrestrial and celestial spheres and the circles that gird the heavens, the equator, the tropics, the ecliptic, and the rest. But when Sacrobosco treated the motions of the ecliptic, he preferred to use the familiar names of the twelve signs that divide it, while Grosseteste had preferred to speak of angles and degrees. Sacrobosco, however, contrasted the risings of the signs presented by the astronomical poets such as

Lucan, Ovid, and Virgil in astronomical and geometrical discussions of spherical astronomy.... Sacrobosco was not concerned with the niceties of technical astronomical detail. Yet his presentation did illustrate the order of nature, noting, as had Bede before him, that the solar eclipse at the Crucifixion must have been miraculous: "Either the God of nature suffers, or the *machina mundi* is dissolved."

Sacrobosco's *De sphaera* is divided into four chapters, the first of which discusses the spherical shape of the universe, the revolution of the heavens, and the measurements of the earth's sphere. The second chapter defines the various circles used to describe the earth and the heavens, including the terrestrial and celestial equators and their respective poles, the meridian, the horizon, the ecliptic, the zodiac circle, the arctic and antarctic circles, and the circles that define the other climatic zones. Chapter 3 gives the coordinates of the zodiacal constellations and explains their heliacal and other risings, as well as explaining the variation of the length of the day in the different global zones. The last chapter describes the movements of the sun, moon, and planets, as well as explaining the causes of lunar and solar eclipses.

De sphaera of Sacrobosco was extremely popular and from the mid-thirteenth century it was taught at universities throughout Europe, continuing in use as a basic text until the end of the seventeenth century. After Manilius's *Astronomica*, it was the first printed book on astronomy, published in Ferrara in 1472. Sixty-four more additions were published in the next seventy-five years, the last being issued in Leiden in 1547, often appearing along with commentaries by distinguished scholars.

Sacrobosco's *De computo eccliastico* deals with the divisions of time into days, months, seasons, and years through the movements of the sun and moon. He pointed out the errors in the Julian calendar, proposing a solution very similar to the reform adopted by Pope Gregory XIII three and a half centuries later. His *Algorismo vulgaris*, or *Common Algorism*, which taught the techniques of calculating with positive integers, was the most widely used manual of arithmetic in the medieval era, continuing in use until the sixteenth century.

Treatises such as Sacrobosco's *De sphaera* only briefly mentioned the epicyclic theory that forms the basis for Ptolemy's *Almagest*, which was studied at European universities beginning in the thirteenth century. Thus most manuscripts of *De sphaera* were supplemented by a section called *Theorica planetarum*, an elementary geometrical description of the epicycle theory of planetary motion. As McCluskey noted:

> The *Theorica planetarum* texts did not go deeply into the trigonometry needed to derive quantitative predictions from these descriptive models; the production of astronomical tables, as opposed to their use, was a specialized matter that did not arise in the introductory course in astronomy. Even astronomical tables, however, ultimately became part of the curriculum, first appearing in the *corpus astronomicum* in the fourteenth century.

A much lengthier and more detailed *theorica planetarum* was written between 1261 and 1264 by Campanus of Novara, who flourished in the second half of the thirteenth century. Campanus is best known for his Latin edition of Euclid's *Elements*, but he was also a distinguished astronomer and mathematician. Little is known of his life other than that he was chaplain to popes Urban IV, Nicholas IV, and Boniface VIII, that in 1270 he was parson of Felmersham in Bedfordshire, in England, and that he spent his last years at the Augustine monastery in Viterbo, in Italy, where he died in 1296.

The principal astronomical work of Campanus is his *Theorica planetarum*, which included a calculation of the dimensions of the universe according to Ptolemy's physical model of the celestial bodies moving in nested crystalline spheres. He based his calculations on the work of the ninth-century Arabic astronomer al-Farghani, who had in turn derived them from the *Planetary Hypotheses* of Ptolemy. Campanus knew of al-Farghani's work from the Latin translation by John of Seville.

Historian Albert van Helden provided a concise summary of the values found by Campanus for the planetary distances, given in miles as well as earth radii (e.r.).

> Using al-Farghani's parameter, he recalculated the planetary distances in miles, adding (as did al-'Urdi) the planetary diameters to the thickness of the spheres. Campanus also used the more precise value of 3,245.5 miles for the Earth's radius. Consequently, he arrived at distances that were slightly different from those of al-Farghani. Saturn's greatest distance, for instance, came out to be 73,387,747 miles, that is 22,612 e.r., instead of al-Farghani's 20,110 e.r. [This was also considered to be the distance to the inner sphere of the fixed stars.]

Campanus's *Theorica planetarum* also provided instructions for the construction and use of an astronomical instrument called the equatorium. This is a scale model of the Ptolemaic epicyclic planetary theory, serving as a mechanical computer for converting the mean position of a planet, found in the tables, to the planet's true position. Campanus probably learned about the equatorium from an Arabic source, for descriptions of equatoria were written nearly two centuries earlier in al-Andalus by ibn al-Samh and al-Zarqali. Through Campanus, more compact and sophisticated versions of the equatorium made their way into the *corpus astronomicum* in the following century, providing a simpler means of computing planetary positions.

Although Campanus intended the equatorium for those who wanted to use it instead of working with conventional astronomical texts, his *Theorica planetarum* also gives detailed instructions in the use of such tables. His examples are obviously taken from the famous *Toledan Tables*. These tables, an adaptation of earlier works by Ptolemy, al-Khwarizmi, and al-Battani, were compiled in Toledo around 1069 by a group of Arabic astronomers, the best known of whom is al-Zarqali. The *Toledan Tables* were used in both al-Andalus and in Christian Europe, where they were translated into Latin in around 1140 as the *Marseilles Tables*. They remained in use until the fourteenth century, and a Latin version of the *Toledan Tables* was translated into Greek, completing a remarkable cultural cycle. The tables are mentioned by Chaucer in "The Franklin's Tale," where one of the characters is a magician-astrologer of Orléans, equipped with all the tools of his celestial trade:

> His tables Toletanes forth he brought
> Ful wel corrected, ne ther lacked noght,
> Neither his collect ne his expans yeres,
> Ne his rotes ne his othere geres....

Although Campanus omitted Ptolemy's mathematical proofs, his *Theorica planetarum* is a highly technical work, used almost exclusively by professional astronomers. He later wrote a more popular work on the sphere called *Tractatus de sphera*, similar to the earlier works of the same type by Grosseteste and Sacrobosco.

Campanus's Latin edition of Euclid's *Elements* is in fifteen books, including the non-Euclidian books 14 and 15. His text is the *editio princeps* of Euclid, and it was reprinted at least thirteen more times in the fifteenth and sixteenth centuries. It was the version in which Euclid was usually studied in the later Middle Ages, remaining in use for three centuries.

Meanwhile, a few translations were still being made from Arabic into Latin in the thirteenth century. Some of these were done under the patronage of King Alphonso X (1221–1284) of Castile and Leon, known in Spanish as el Sabio, or the Wise. Alfonso's active interest in science led him to sponsor translations of Arabic works in astronomy and astrology, including a new edition of the *Toledan Tables*. This edition, known as the *Alphonsine Tables*, included some new observations but retained the Ptolemaic system of eccentrics and epicycles.

One of the most notable of the early European astronomers was William of St. Cloud, who flourished in France during the late thirteenth century. The earliest date of his activity is December 28, 1285, when he observed a conjunction of Saturn and Jupiter, to which he refers in his *Almanach*, completed in 1292. His *Calendrier de la reine*, also completed in 1292, proved to be his most lasting contribution. He dedicated the book to Queen Marie of Brabant, widow of Philip III the Bold, and translated it into French at the request of Jeanne of Navarre, wife of Philip IV the Fair.

Queen Marie's *Calendrier* represents William's effort to put the calendar on a purely astronomical basis. This led him to contradict the computations in the ecclesiastical calendar, which he found full of errors, indicating the need for calendar reform. Likewise, his *Almanach* provided listings in which the positions of the celestial bodies were given directly, as contrasted to earlier tables, which only gave the elements by which those positions could be calculated.

He points out the errors in the earlier planetary tables and shows how he corrected them. These tables were those of Toledo, used in the Muslim calendar, and of Toulouse, the adaptation of the *Toledan Tables* to the Christian calendar. William's *Almanach* makes no mention of the *Alphonsine Tables*, which were not used in Paris before 1320. During the period 1285–1292 he observed the sun, moon, Mars, Jupiter, and Saturn in order to check and improve the accuracy of astronomical tables. He used his observations of the sun to measure the obliquity of the ecliptic as 23°34', within 2 minutes of arc of the modern value for that era. His estimation of the time of the spring equinox for 1290 was within six hours of the values from modern tables, a great improvement over the values given in ecclesiastical calendars of that era. He used his observations to resolve the question of the slow movement of the outer sphere of the fixed stars, rejecting the trepidation theory in favor of the precession of the equinoxes, where he chose al-Battani's value of 1° in sixty years instead of Ptolemy's estimate of 1° in somewhat less than a century, the actual rate being 1° in seventy-two years. He also observed conjunctions of Saturn with Jupiter, and of Mars with the moon and with a star in the constellation Scorpio. He then used these observations to correct values of the mean motions in the *Toulouse Tables*, incorporating these into his *Almanach*.

The *Calendrier* refers to a treatise on the *Directorium*, or "adrescoir," the earliest of William's extant works. According to French historian

Emmanuel Poulle, "The instrument described in it is a magnetic compass with a graduation in unequal hours; it is provided with a table for computing the duration of diurnal arc."

The vexing question of the movement of the sphere of fixed stars was also addressed by John of Sicily, who was a member of the Paris scientific community at the end of the thirteenth century. His only extant work is a commentary, dated 1290 or 1291, of Gerard of Cremona's translation of the canons of al-Zarqali's planetary tables. The commentary has the character of a treatise on trigonometry and planetary astronomy. It is divided into three parts, the first of which discusses what Gerard of Cremona called the measure of time, that is, matters concerning the divisions of the year in the lunar and solar calendars, and the conversion of dates between the Christian, Islamic, Greek, and Persian calendars.

The version of the *Toledan Tables* used in this canon reduced the years in all of the calendars to days in sexagesimal notation, a principle later adopted and systematized in the *Alphonsine Tables*.

The second part of the book deals with the astronomy of the *primum mobile*, or prime mover, the outer sphere of the fixed stars, and includes the application of trigonometry to astronomy. Here John rejected the trepidation theory in favor of the precession of the equinoxes, stating that the exact value for the rate of precession was uncertain and would only become definite after more extensive observation. The third and last part is concerned with the motion of the planets.

A new set of astronomical tables was completed in 1327 by John of Lignères and his students, whose work was heavily dependent on the *Toledan Tables*. This work, known as the *Large Tables*, which included a catalogue of the positions of the forty-seven brightest stars, was easier to use than any of the earlier tables and thus became very popular, though it was eventually supplanted in Paris by the *Alphonsine Tables*.

John of Saxony, a student of John of Lignères, published canons on the *Alphonsine Tables* in 1327. These canons are clearer and more extensive than those of John of Lignères, with added chapters on finding the time for the sun to return to a previously occupied position; calculating the date and hour of a true conjunction of the sun and moon; determining the time when the sun enters one of the signs of the zodiac; and establishing the dates of planetary conjunctions. John of Saxony's canons became very popular, as evidenced by the number of extant manuscript and by their inclusion in the first printed edition of the *Alfonsine Tables*, published in 1483. He also wrote a commentary on the popular astrological treatise of al-Qabisi, the third Latin edition of which was published together with the commentary in 1475.

Levi ben Gerson (1288–1344) was a Jewish polymath who wrote books on astronomy, physics, mathematics, and philosophy, as well as commentaries on the Bible and the Talmud. He lived in Orange and Avignon, which were not affected by the expulsion of the Jews from France in 1306 by King Philip the Fair. He was also on good terms with the papal court in Avignon, as evidenced by his dedication of one of his works to Pope Clement IV in 1342.

Levi's greatest work is his *Milhamot Adonai* (*The Wars of the Lord*), a philosophical treatise in six books, the fifth of which is devoted to astronomy. Here Levi presents his model of the universe, based on several Arabic sources, principally al-Battani, Jabir ibn Aflah, and ibn Rushd (Averroës). His model differed in important respects from that of Ptolemy, whose theories did not always agree with observations made by Levi. This was particularly so in the case of Mars, where Ptolemy's theory had the apparent size of the planet varying by a factor of six, whereas Levi's observation found that it only doubled.

The instruments used by Levi included one of his own invention, the so-called Jacob's Staff, a device to measure angles in astronomical observations. He also employed the camera obscura, which he used in observing eclipses and in determining the eccentricity of the sun's orbit. Levi's astronomical work was influential in Europe for five centuries, and his Jacob's Staff was also used for maritime navigation until the mid-eighteenth century.

His *Sefer ha-mispar* (*The Book of the Number*) deals with general principles of arithmetic and algebra and their applications to combinatorial analysis and summations of series. The importance of this work is that here Levi seems to have been the first to use mathematical induction in his demonstrations, before Francesco Maurolyco (1575) and Blaise Pascal (c. 1654).

During the first half of the fourteenth century an important school of astronomy was active at Oxford. The most notable of the Oxford astronomers was Richard of Wallingford (c. 1292–1336). Richard, the son of a blacksmith, studied at Oxford from c. 1308 until 1315, when he joined the Benedictine order at St. Alban's abbey. Two years later he was sent back to Oxford for further studies, remaining there until 1327, when he was appointed abbot of St. Albans. After his appointment he visited Avignon for the papal confirmation, and when he returned to St. Albans he found that he had contracted leprosy, from which he died in 1336.

Richard's works were all written during his years at Oxford. One of these was his *Quadripartitium*, the first comprehensive treatise on spherical trigonometry written in Latin Europe. Richard's most important work was his *Tractus albionis*, which dealt with the theory, construction, and use of an instrument that he had invented called the Albion ("all by one"), a form of equatorium used to perform all sorts of astronomical measurements and calculations, used throughout Europe up until the sixteenth century.

While writing the *Tractus albionis* Richard composed a treatise on another new astronomical instrument he had designed. This was the *rectangulus*, designed as a simpler substitute for the armillary sphere. The treatise also contained a table of the inverse trigonometric function for use in the analysis of the observations with the *rectangulus*.

Richard also built an enormous mechanical clock, ten feet in diameter, which he installed on the wall of the south transept in the abbey church. Besides the time of day, it also showed the motion of the celestial bodies, the phases of the moon, and the tides under London Bridge. The clock was destroyed in the sixteenth century, but several drafts of Richard's design have survived, the oldest extant plans of any mechanical clock, one that was the most sophisticated of the medieval era. (The earliest mechanical clocks in Europe appear to date from the end of the thirteenth century.) The mechanical clock led to the notion of time as a physical quantity that could be expressed numerically in terms of units on a scale and used in scientific theories. Since Richard's clock was also a planetarium, it lent credence to the notion that the universe was a divinely designed clockwork.

The earliest appearance of a mechanical clock in literature seems to be a passage of Dante's *Paradiso*, in the last lines of canto 10, written between 1316 and 1321, a decade or so before Richard built his clock.

> Forthwith
> As clock, that calleth up the spouse of God
> To win her Bridegroom's love at matin's hour.
> Each part of other fitly drawn and urged,
> Sends out a tinkling sound, of note so sweet,
> Affection springs in well-disposed breast;
> Thus saw I move the glorious wheel; thus heard
> Voice answering voice, so musical and soft,
> It can be known but where day endless shines.

The brilliant period of scholastic thought that produced these gifted scholars came to a sudden end toward the close of the fourteenth century, and for the next century and a half Oxford and Paris produced no scientific work of any originality or importance.

This hiatus was due to the plague known as the Black Death, which ravaged Europe in the years 1347–1350 and recurred intermittently for centuries, its victims including some of the Continent's leading scholars. As Lindberg noted, writing of the primitive state of medical science at the time, when astrological influence was frequently given as a cause of major epidemics: "Pressed for an explanation of this particular plague, the medical faculty at the University of Paris concluded that it resulted from a corruption of the air caused by a conjunction of Jupiter, Saturn, and Mars in 1345."

Nicholas of Cusa (1401–1464), noted for his revolutionary ideas about science and cosmology, was the first major intellectual figure to emerge after Europe recovered from the Black Death. Born in Cues, a village on the Moselle, he was educated at the universities of Heidelberg and Padua, where he received a doctorate in 1423. He entered the priesthood in around 1430 and in 1448 Pope Nicholas V made him a full cardinal, with the titular see of St. Peter in Vincoli in Rome.

Nicholas of Cusa

Cusa's most important work is his *De docta ignorantia* (*On Learned Ignorance*), completed in 1440. Here he uses mathematics and experimental science in his attempts to determine the limits of human knowledge, particularly the inability of the human mind to conceive the absolute, which to him was the same as mathematical infinity. He concluded that the universe was infinite in extent, making the idea of a center or of a periphery meaningless. Thus the earth cannot be the center of the universe, and since motion is relative and natural to all bodies, the earth cannot be at rest. A marginal note made by Cusa in one of his manuscripts suggests that the earth cannot be fixed, but rotates on its axis once in a day and a night.

Cusa speculated that the earth might not be the only body on which there were living creatures, and that there might be another earth at the center of the sun's luminous envelope. These and other revolutionary theories led his political rivals to accuse him of pantheism, a charge that he defended himself against in his *Apologia doctae ignorantia* (1440), in which he quoted patristic writings and Christian Neoplatonists as the sources of his ideas. The criticism of Cusa's ideas foreshadows Galileo's troubles with the Church two centuries later, but in mid-fifteenth century the situation was not as highly charged as it would be a century later, after the beginning of the Protestant Reformation.

Ten years later Cusa wrote a work entitled *Idiota*, a series of dialogues in which a rhetor, or schoolman, who represents book learning, converses with a layman (Idiota), who stresses the importance of quantitative experimental research. According to Idiota, wisdom clamors in the streets and one can find it in the marketplace, where one sees money being counted, merchandise being weighed, oil being measured, and where one can watch human reason performing its most fundamental function: measurement.

While western Europe was steadily advancing intellectually, it remained virtually cut off from the Byzantine Empire, at least in the early medieval period. Soon after Constantine the Great established his new capital at Constantinople in 330, he founded a university there. Nothing is known about the original University of Constantinople, which may not actually have functioned until it was reorganized in 425 by Theodosius II. The new university originally had twenty chairs of grammar, equally divided between Greek and Latin, and eight in rhetoric, five of which were in Greek and three in Latin, as well as two professorships in law, one each in Greek and Latin, and one in philosophy, taught in Greek. By the following century Latin had fallen out of use in Constantinople and all of the professorships at the university were in Greek. This was part of the great cultural divide that developed in the early medieval era between Latin West and Greek East, a dichotomy that separated the newly emerging civilization of western Europe from the Byzantine world of the Balkans and Asia Minor.

The Byzantine Empire almost came to an end in 1204, when Constantinople was captured and sacked by the Latin knights of the Fourth Crusade, with the aid of the Venetians under the command of Doge Enrico

Dandolo. The Latins ruled in Constantinople until 1261, when the city was recaptured by the Greeks under the emperor Michael VIII Palaeologus (r. 1259–1282), founder of the dynasty that ruled Byzantium until its conquest by the Turks in 1453.

Byzantium flowered in a last renaissance in the Palaeologan era, as evidenced by the extraordinary paintings and frescoes in the early-fourteenth-century church of St. Saviour in Chora (Kariye Cami), preserved in Istanbul.

One aspect of this renaissance was a revival of astronomy, which began at the end of the thirteenth century when Greek translations of Arabic astronomical treatises became available in Constantinople. One of the first such translations was done by Gregory Choniades, who learned Arabic science in Tabriz and founded a school of astronomy in Trebizond around 1300. By the mid-fourteenth century new *Persian Tables* were substituted for the older astronomical tables, as in the *Three Books on Astronomy*, a compendium written around 1352 by Theodore Meliteniotes, who became head of the Patriarchal School in Constantinople, a revival of the university that had been founded by Constantine the Great.

The fall of Thessalonica to the Ottomans in 1430 led the emperor John VIII Palaeologus (r. 1425–1448) to seek help from the West, and he proposed to Pope Martin that a council be called to help reconcile the Greek and Latin churches. This eventually gave rise to a council that was convened by Pope Eugenius IV in 1438 in Ferrara, moving to Florence the following year. The council ended on July 6, 1439, when a Decree of Union between the Greek and Latin churches was read in Latin and Greek in the cathedral of Florence, Santa Maria del Fiore, in the presence of the emperor John VIII. But most of the people and clergy of Byzantium rejected the Union, dividing the empire in what were to be the last years of its existence.

The delegates to the Council of Ferrara-Florence included four scholars who were leading figures in the cultural interchange between Byzantium and Italy, one of them representing the Roman Catholic Church and the others the Greek Orthodox Church. The Catholic delegate was Nicholas of

Cusa; the Greeks were George Gemistus Plethon, George Trapezuntios, and Basilios Bessarion.

George Gemistus Plethon (c. 1355–1452), whom historian Sir Steven Runciman called "the most original of all Byzantine thinkers," was educated in Constantinople and taught there until about 1392. He then went to Mistra in the Peloponnesus, which at the time was ruled by the despot Theodore Palaeologus, second son of the emperor Manuel II (r. 1391–1425). Plethon taught there for the rest of his days, except for a year that he spent as a member of the Byzantine delegation at the Council of Ferrara-Florence. His teaching was dominated by his rejection of Aristotle and his devotion to Plato, who inspired his goal of reforming the Greek world along Platonic lines. His religious beliefs were more pagan than Christian, as evidenced by his treatise *On the Laws*, in which he usually refers to God as Zeus and writes of the Trinity as consisting of the Creator, the World-Mind, and the World-Soul. George Trapezuntios wrote of a conversation he had in Florence with Plethon, who told him that the whole world would soon adopt a new religion. When asked if the new religion would be Christian or Mohammedan, Plethon replied: "Neither, it will not be different from paganism."

While the council was deliberating in Florence, Plethon delivered a lecture at the palace of Duke Cosimo de' Medici, the subject being the philosophical and religious differences between Platonism and Aristotelianism, in which he eulogized Plato. Cosimo was so inspired by Plethon that he founded a Platonic Academy in Florence, which became the center of renaissance Platonism. Plethon's writings also inspired Masilio Ficino (1433–1499), the first of the great Renaissance Platonists.

George Trapezuntios (1395–1486) was born on Crete to a family who had moved there from Trebizond, hence his last name. He was a prolific translator from Greek into Latin, which he studied in Venice in 1417–1418. He taught in Venice, Vicenza, and Mantua before coming to Rome, where he served in the papal bureaucracy under Eugenius IV (r. 1431–1447) and then taught under Nicholas V (1447–1465). He severely criticized Plethon for his attack on Aristotelianism, portraying himself as the champion of medieval scholasticism against its humanist critics, and singling out for special praise Albertus Magnus and Thomas Aquinas, which brought him into conflict with Bessarion, among others.

Basilios Bessarion (c. 1403–1470) was born into a family of manual laborers in Trebizond. The metropolitan of Trebizond noticed the boy's intelligence and sent him to school in Constantinople. While at the University of Constantinople he met the Italian humanist Francesco Filelfo, who had been inspired to study there after attending the classes of the Byzantine scholar Manuel Chrysoloras in Italy.

At the age of twenty Bessarion became a monk and spent some years at a monastery near Mistra, where he studied under Plethon. He then returned to Constantinople and won renown as a professor of philosophy. He was chosen as one of the delegates to the Council of Ferrara-Florence, and was appointed metropolitan of Nicaea so that he would have appropriate status at the conclave. When the agreement of union was formally proclaimed on July 6, 1439, in the cathedral in Florence, it was first read in Latin by Cardinal Cesarini and then in Greek by Bessarion.

Bessarion's stay in Italy convinced him that Byzantium could only survive in alliance with the West, not just politically, but also by sharing in the cultural life of Renaissance Italy. Disheartened by opposition to the Union in Constantinople, he returned to Florence at the end of 1440, by which time he had already been made a cardinal in the Roman Catholic Church. He spent some time traveling on papal diplomatic missions and served as governor of Bologna from 1450 to 1455, but otherwise from 1443 on he resided in Rome. He was nearly elected pope in 1455, but he lost out when his enemies warned of the dangers of choosing a Greek, and so the cardinals turned to the Catalan Alfonso Borgia, who was elevated as Callisto IV.

Much of Bessarion's energy was spent trying to raise military support in Europe to defend Byzantium against the Turks, but his efforts came to naught, as the Ottomans captured Constantinople in 1453 and then took his native Trebizond in 1461, ending the long history of the Byzantine Empire. Thenceforth Bessarion sought to find support for a crusade against the Turks, but to no avail.

Bessarion devoted much of his time to perpetuating the heritage of Byzantine culture by adding to his collection of ancient Greek manuscripts, which he bequeathed to Venice, where they are still preserved in the Marciana Library. The group of scholars who gathered around Bessarion in Rome included George Trapezuntios, whom he commissioned to translate Ptolemy's *Almagest* from Greek into Latin. Then in 1459 Trapezuntios published an attack on Platonism, suggesting that it led to heresy and immorality. Bessarion was outraged and wrote a defense of Platonism, published in both Greek and Latin. His aim was not only to defend Platonism against the charges made by Trapezuntios, but to show that Plato's teachings were closer to Christian doctrine than those of Aristotle. His book was favorably received, for it was the first general introduction to Plato's thought, which at the time was unknown to most Latins, for earlier scholarly works on Platonism had not reached a wide audience.

One of Bessarion's diplomatic missions took him in 1460 to Vienna, whose university had become a center of astronomical and mathematical studies through the work of John of Gmunden (d. 1442), Georg Peurbach (1423–1461), and Johannes Regiomontanus (1436–1476). John had built astronomical instruments and acquired a large collection of manuscripts, all of which he had bequeathed to the university, thus laying the foundations for the work of Peurbach and Regiomontanus.

Peurbach was an Austrian scholar who had received a bachelor's degree at the University of Vienna in 1448 and a master's in 1453, while in the interim he had traveled in France, Germany, Hungary, and Italy. He had served as court astrologer to Ladislaus V, king of Hungary, and then to the king's uncle, the emperor Frederick III. His writings included textbooks on arithmetic, trigonometry, and astronomy, his best-known treatises being his *Theoricae novae planetarum* (*New Theories of the Planets*), as well as two unfinished works that were completed and published by Regiomontanus, the *Tabulae eclipsium* (*Tables of Eclipses*), and the *Epitome Almagesti Ptolemaei* (*Epitome of Ptolemy's Almagest*). His mathematical and astronomical works were well in advance of their time and would be used by Copernicus, among others.

Regiomontanus, originally known as Johann Muller, took his last name from the Latin for his native Königsberg in Franconia. He studied first at the University of Leipzig from 1447 to 1450, and then at the University of Vienna, where he received his bachelor's degree in 1452, when he was only fifteen, and his master's in 1457. He became Peurbach's associate in a research program that included a systematic study of the planets as well as observations of astronomical phenomena such as eclipses and comets.

Bessarion was dissatisfied with the translation of Ptolemy's *Almagest* that had been done by George Trapezuntios, and he asked Peurbach and Regiomontanus to write an abridged version. They agreed to do so, for Peurbach had already begun writing a compendium on the *Almagest*, as well as a table of eclipses, but both were unfinished when he died in April 1461. On his deathbed he made Regiomontanus promise that he would complete both works. Regiomonantus swore that he would, and he finished both of them about a year later in Italy, where he had gone with Bessarion. He spent part of the next four years in the cardinal's entourage and the rest in his own travels, learning Greek and searching for manuscripts of Ptolemy and other ancient astronomers and mathematicians.

De triangulis omnimodis, which Regiomontanus dedicated to Bessarion, is a systematic method for analyzing both plane and spherical triangles using trigonometry. According to historian Edward Rosen, one of the propositions on spherical triangles in this work "solves, for the first time in the Latin West, a trigonometric problem by means of algebra (here called the *ars rei et census*)." He goes on to say that this work, which was not published until 1533, "attracted many important readers and thereby exerted an enormous influence on the later development of trigonometry because it was the first printed systematization of that subject as a branch of mathematics independent of astronomy." Carl B. Boyer went as far as to say that the publication of *De triangulis omnimodis* represents "the rebirth of trigonometry."

The *Tabulae directionum*, which was first published in 1490 and very frequently thereafter, consisted of tables of the longitudes of the celestial bodies in relation to the apparent daily rotation of the heavens. Here Regiomontanus included a table of the tangent function (using the word *numerus*), which he had not given in *De triangulis omnimodis*. The tables that he gave for the sine and tangent functions provided the model for those used in modern trigonometry textbooks.

Regiomontanus left Italy in 1467 for Hungary, where he served for four years in the court of King Mathias Corvinus, continuing his researches in astronomy and mathematics. Then in 1471 he moved to Nuremberg, where he set up his own observatory and printing press, which he operated until his premature death in 1476.

The first book he printed was Peurbach's *Theoricae novae planetarum*. The *Theorica* became extremely popular, going through fifty-six editions by the middle of the seventeenth century. The great importance stemmed from the fact that it gave careful and detailed descriptions of Ptolemy's planetary theory of crystalline spheres, which remained the accepted model of celestial structure until the late sixteenth century, when Tycho Brahe showed that the planetary spheres did not exist. It influenced astronomers up to the time of Copernicus, who was led to his own planetary theory through his attempt to correct Peurbach's models.

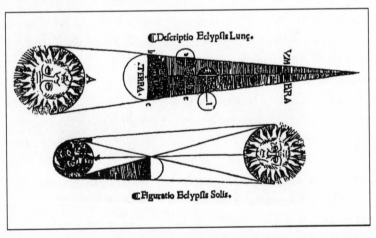

Diagram of the lunar and solar eclipses from Peurbach's
Theoricae novae planetarum, *1525*

From the Egyptoma in Almagestum *(1496) by Regiomontanus, showing Ptolemy (left) and Regiomontanus (right) below an armillary sphere*

The second book published by Regiomontanus was his own *Ephemerides*, the first planetary tables ever printed, giving the positions of the heavenly bodies for every day from 1475 to 1506. Columbus is said to

· 213

have taken the *Ephemerides* with him on his fourth and last voyage to the New World, and to have used its prediction of the lunar eclipse of February 29, 1504, to frighten the hostile natives of Jamaica into submission.

Regiomontanus also published the two works of Peurbach that he had completed after the death of his colleague, the *Tabulae eclipsium* and the *Epitome Almagesti Ptolemaei*.

The *Tabulae eclipsium* was completed probably in 1459. It is based on the *Alphonsine Tables*, but, as historians Doris Hellman and Noel Swerdlow pointed out, "Peurbach expanded and rearranged the tables needed for every step in eclipse computation, saving the calculator much time and relieving him of a number of tedious procedures." The *Tabulae eclipsium* continued in use until the end of the sixteenth century, when Tycho Brahe employed it despite his criticism of some of its contents.

The *Epitome*, which was dedicated to Bessarion, was not published until 1496, twenty years after the death of Regiomontanus. The work was not a mere abridged translation of the *Almagest*, as Edward Rosen pointed out, for "it added later observations, revised computations, and critical reflections—one of which revealed that Ptolemy's lunar theory required the apparent diameter of the moon to vary in length much more than it really does." This particular passage came to the attention of Copernicus, who first read the *Epitome* when he was a student in Bologna, and may have been one of the factors that led him to replace Ptolemy's geocentric theory with his own planetary model, a revival of the heliocentric theory first proposed by Aristarchus of Samos.

At least two other propositions in the *Epitome* influenced Copernicus in the formulation of his own planetary theory. These propositions seem to have originated with the fifteenth-century Arabic astronomer Ali Qushji and may have been transmitted to Regiomontanus by Bessarion. If so, this would place Bessarion and Regiomontanus in the long chain that leads from Aristarchus of Samos and Ptolemy to Copernicus through the Arabic and Latin scholars of the Middle Ages to the dawn of the Renaissance.

The confluence of ideas that we have seen flowing into western Europe, from Islam as well as Byzantium, is evidence that the Renaissance was dawning. Islamic science, though now past its prime, continued to produce original works in mathematical astronomy, which made their way to the West principally through Byzantine scholars in Trebizond. As we have seen, Byzantium played an important part in the Italian Renaissance prior to the conquest of Constantinople in 1453, and it continued to do so even afterward through the efforts of Cardinal Bessarion of Trebizond. Bessarion was the central figure in this cultural interchange, having bridged the gap between Greek East and Latin West that had persisted for more than a thousand years.

I thought of this when visiting Bessarion's tomb in Rome a few years ago, in a portico of the Conventual Minorities, close to the Basilica of the Twelve Apostles. For me, Bessarion represents the end of a tradition that began with Thales of Miletus, transmitting the spirit of the Ionian enlightenment directly to Copernicus and those who followed him. This includes my humble self, for my scientific genealogy takes me all the way back to Bessarion, who in 1423 graduated from the University of Constantinople, today's Istanbul, where I am still teaching, as are some of my students and their students.

11

The Revolutions of the
Celestial Spheres

*T*HE REVIVAL OF ASTRONOMY CULMINATED WITH THE PARADIGM-shifting work of Nicolaus Copernicus (1473–1543), whose life spanned the transition period between the twilight of the medieval era and the dawn of the Renaissance. He was born twenty years after the fall of Constantinople, capital of the Byzantine Empire, the Christian continuation of the Roman Empire. Two years later the Gutenberg Bible was printed, stimulating an unprecedented spread of the new learning that had developed in western Europe. Copernicus's teenage years saw monumental changes such as Columbus's discovery of America, opening up a new world at the beginning of a new age.

Copernicus was born on February 19, 1473, in Torun, a town on the Vistula 110 miles northwest of Warsaw, in what was then the Duchy of Prussia and is now Poland. His name was originally Niklas Koppernigk, which he Latinized as Nicolaus Copernicus after he went to university. The youngest of four children of a prosperous merchant, his father died in 1483, whereupon he and his siblings were adopted by their maternal uncle Lucas Watzenrode, a priest who had studied at the universities of Cracow, Cologne, and Bologna.

In 1489 Lucas become bishop of Ermland, also known as Warmia, one of the four provinces into which the Duchy of Prussia was then divided, with the kingdom of Prussia to its west and the kingdom of Poland to its south. Copernicus and his older brother, Andreas, stayed with their uncle Lucas at his palace in Heilsburg, 140 miles northeast of Torun, while their sister Barbara entered a convent. Their other sister, Maria, married a merchant in Cracow.

In the autumn of 1491 Nicolaus and Andreas were sent by their uncle Lucas to the University of Cracow, where they enrolled in the faculty of arts. They remained there for about three or four years, but left without taking a degree. During that time Nicolaus is known to have taken courses in mathematics, astronomy, astrology, and geography. According to his first biographer, the French astronomer Pierre Gassendi (1592–1655), his reading also included Cicero, Virgil, Ovid, and Seneca, for the curriculum at Cracow was in tune with the spirit of the Renaissance, where the emphasis was on the humanities rather than science.

The renowned Polish astronomer Albert Brudzewski was lecturing in the University of Cracow at the time and Nicolaus would undoubtedly have read his works, although there is no record of their having met. Brudzewski had published a commentary on the *Theorica novae planetarum* of Peurbach, in which he put forward his own theory that the celestial orbs are not spheres but circles. Brudzewski also used a mathematical method analogous to one employed by the Arabic astronomers Nasir al-Din al-Tusi and ibn al-Shatir, similar to a model that Copernicus would later use in his heliocentric theory. Brudzewski may have obtained this theory from a work written in Cracow in 1430 by the Polish astronomer Sandivogius of Czecel. This suggests that the works of Nasir al-Din al-Tusi and ibn al-Shatir were available at the University of Cracow when Copernicus was studying there.

According to Gassendi, the textbooks that Copernicus read in his courses in mathematics, astronomy, and astrology included works by Euclid, Ptolemy, Sacrobosco, Peurbach, and Regiomontanus. In addition, the works of a number of Arabic astronomers were available in Cracow at that time, including those of al-Farghani, al-Kindi, Thabit ibn Qurra, al-Tusi, and al-Shatir. Copernicus also frequented Johann Haller's bookshop in Cracow, where he purchased the *Alphonsine Tables* and the *Tabulae directionum* of Regiomontanus and had them bound together with parts of Peurbach's *Tables of Eclipses* and tables of planetary latitudes.

Nicolaus and Andreas left Cracow early in 1496 to live with their uncle Lucas in the bishop's palace at Heilsburg. Lucas nominated Nicholas and Andreas to be canons of Frauenburg Cathedral, that is, resident clergy who were not required to take holy orders or a vow of celibacy, but at first his efforts were unsuccessful. Nicolaus was finally made a canon in 1497 and Andreas was elected in 1499. Both of them were elected in absentia, for in the fall of 1496 Nicolaus had gone off to study at the University of Bologna, where Andreas joined him two years later. There, both joined the faculty of law and enrolled in the *Natio Germanorum*, the largest of the "nations" into which foreign students were organized at Bologna.

The brothers seem to have stayed as paying guests in the house of Domenico Maria da Novara (1454–1504) of Ferrara, a professor of astronomy at the university. Nicolaus believed that he was "not so much the pupil as the assistant and witness of observations of the learned Dominicus Maria," as his friend Rheticus later quotes Copernicus. One of the observations in Bologna concerned a lunar occultation of the star Aldabaran, which Copernicus says they made "after sunset on the seventh day before the Ides of March, in the year of Christ 1497."

Copernicus received his master's degree in law at Bologna in 1499. He then went to Rome early in 1500 to take part in the celebration of the Jubilee, or Holy Year, that had been proclaimed by Pope Alexander VI. According to Rheticus, while in Rome Copernicus "lectured on mathematics before a large audience of students and a throng of great men and experts in this branch of knowledge." Copernicus notes in *De revolutionibus* that he observed a lunar eclipse in Rome on November 6, 1500. He compared it to an eclipse observed by Ptolemy in Alexandria in the "nineteenth year of Hadrian" (AD 136/137), his purpose being "to determine the positions of the moon's movement in relation to the established beginnings of calendar years."

Nicolaus and Andreas returned to Poland in May 1501. On July 27 of that year they made an appeal to the authorities of their chapter in Frauenburg, asking for a two-year extension of their leave so that they could complete their studies in Italy. The chapter accepted and in August they left Frauenburg for Italy, Andreas to complete his degree in canon law in Bologna and Nicolaus to study medicine in Padua.

Nicolaus enrolled at the University of Padua in the fall of 1501, study-ing law as well as medicine. He interrupted his studies in Padua after two years to enroll at the University of Ferrara, where on May 31, 1503, he re-ceived the degree of doctor of canon law. Copernicus practiced medicine for the rest of his career in addition to continuing his work in astronomy. He was truly a Renaissance man before the term was coined, for to him and other scholars of his time there were no boundaries between the aca-demic disciplines, and he brought the same attitude of mind to his medical work as he did to his astronomical researches.

Copernicus returned to Poland later in 1503, first rejoining his fellow canons at Frauenberg and soon afterward going on to live with his uncle Lucas at Heilsburg Castle (Lidzbark Warminski), the official residence of the bishops of Warmia, about forty miles southeast of Frauenburg. He re-mained at Heilsburg for six years, serving as secretary of state and personal physician to his uncle Lucas. Copernicus also participated in the local Pruss-ian diets, or parliaments, as a member of the group of canons called the Chapter of Warmia.

On January 1, 1504, Copernicus went to Marienburg (Malbork) to attend a meeting of the Land Diet of the Prussian States, at which Bishop Lucas presided. The assembly decided to convene another meeting twenty days later in Elbing (Elblag), where the Prussian states solemnly refused to make a pledge of loyalty to King Alexander, grandson of Wladyslaw II Jagiello, who had become Grand Duke of Lithuania in 1492 and succeeded his childless elder brother, John I Albert, as king of Poland in 1501, reunit-ing the two states.

In May 1504 Nicolaus went with Lucas to a meeting of the Prussian delegates with King Alexander in Thorn, after which they accompanied the king to Danzig. He attended another assembly of the Prussian states with his uncle in 1506 from August 20, to September 15. Nicolaus proba-bly also accompanied Lucas to Cracow on January 24, 1507, to attend the coronation of Sigismund I, who had succeeded his brother Alexander as king of Poland the previous year.

Nicolaus Copernicus, from the 1554 Paris edition
of his biography by Pierre Gassendi

On January 24, 1507, Nicolaus was appointed personal physician of the bishop of Warmia, with a salary of 15 marks a year in addition to his income as a canon, which in 1519 was 98 marks. Soon afterward Bishop Lucas became seriously ill, but Nicolaus nursed him back to health, and he accompanied his uncle to an assembly of the Prussian states from September 1 through September 4 of that year, the last one they would attend together.

Around 1509 Nicolaus left his uncle's service in Heilsberg and rejoined the cathedral chapter in Frauenburg, where he would spend most of the remainder of his life. It may be that by then he had decided to return to the astronomical research he had begun in Italy, which he would not have had the time to do while serving as his uncle's minister of state.

Early in 1512 he accompanied his uncle to Cracow to attend the wedding of King Sigismund and the coronation of his bride. But on the way home Bishop Lucas died in Torun, on March 29, 1512. His body was brought back to Frauenburg and, as was the custom at the time, he was buried in the cathedral, where his tomb can still be seen, with an epitaph composed by Copernicus.

When Copernicus returned to Frauenburg, he was one of sixteen canons in the cathedral chapter of Warmia. Most of the canons lived in a dormitory beside the cathedral. Each of them also had a curia, a house in the town outside the walls on Cathedral Hill, and some also had a villa in the countryside. In 1514 Copernicus bought a house just outside the west gate of Cathedral Hill. The previous year he had moved out of the dormitory inside the fortress and took up residence within the defense tower in the northwest corner of the fortress walls, which he had purchased from the cathedral chapter. The tower had three floors, the uppermost of which had windows on all sides, so that he could look out over the town and the surrounding countryside. Nearby he also built an observatory in the form of a viewing platform, where he set up his astronomical instruments with an unobstructed view of the celestial sphere. The observatory was built at his own expense, as recorded in the chapter archives, which note that in April 1513, he paid money into the chapter treasury for eight hundred bricks and a barrel of chlorinated lime from the cathedral work yard.

In *De revolutionibus* Copernicus refers to three instruments that he may have used in his observations. One was a sighting device called a *triquetrum*, or Ptolemy's ruler, used for measuring parallax. Another was a quadrant, a kind of sundial set in a wooden block, used to measure the altitude of the sun at noon. The third was an armillary astrolabe, also known as an armillary sphere, which Copernicus describes as a set of nested rings, one set of which contained sights for measuring the position of a star or planet. He undoubtedly also had a regular astrolabe, although he does not mention it in *De revolutionibus*.

Looking up from his tower on June 5, 1512, Copernicus noted that Mars was in opposition, that is, the planet rose at sunset and set at sunrise, since it was diametrically opposite the sun in the celestial sphere. This was the first of at least twenty-five observations that Copernicus would make at Frauenburg, where he also developed the mathematical methods that he used in his new astronomical theory.

Around this time Copernicus began writing a work entitled *De hypothesibus motuum coelestium a se constitutis commentariolus* (*Sketch of Hypotheses for the Celestial Motions*). This came to be known as the *Commentariolus*, or "Little Commentary," the first notice of the new astronomical theory that Copernicus had been developing. He gave written copies of this short treatise to a few friends but never published it in book form. Only two manuscript copies have survived, one of which was first published in Vienna in 1878. The earliest record of the *Commentariolus* is a note made in May 1514 by a Cracow professor, Matthias de Miechow, who writes that he had in his library "a manuscript of six leaves expounding the theory of an author who asserts that the earth moves while the sun stands still." He was unable to identify the author of this treatise, since Copernicus, with his customary caution, had not written his name on the manuscript. But there is no doubt that the manuscript was by Copernicus, because the author made a marginal note that he had reduced all his calculations "to the meridian of Cracow, because ... Frombork [Frauenburg] ... where I made most of my observations ... is on this meridian as I infer from lunar and solar eclipses observed at the same time in both places." (Frombork is actually about ¼° west of Cracow.)

The introduction to the *Commentariolus* discusses the theories of Greek astronomers concerning "the apparent motion of the planets," noting that the homocentric spheres of Eudoxus were "unable to account for all the planetary motions" and were supplanted by Ptolemy's "eccentrics and epicycles, a system which most scholars finally accepted." But Copernicus took exception to Ptolemy's use of the equant, which led him to think of formulating his own planetary theory, in which "the center of the earth is not the center of the universe." He said that "we revolve around the sun like any other planet" and that the apparent daily rotation of the heavens results from the real diurnal rotation of the earth in the opposite sense.

Copernicus went on to say that after setting out to solve "this very difficult and almost insoluble problem," he finally arrived at a solution that involved "fewer and much simpler constructions than were formerly used," provided that he could make certain assumptions.

The assumptions, seven in number, were: (1) that there is not a single center for all the celestial circles, or spheres; (2) that the earth is not the center of the universe, but only of its own gravity and of the lunar sphere; (3) that the sun is the center of all the planetary spheres and of the universe; (4) that the earth's radius is negligible compared to its distance from the sun, which in turn is "imperceptible in comparison to the height of the firmament"; (5) that the apparent diurnal motion of the stellar sphere is due to the rotation of the earth on its axis; (6) that the daily motion of the sun is due to the combined effect of the earth's rotation and its revolution around the sun; and (7) that "the apparent retrograde and direct motion of the planets arise not from their motion but from the earth's." He then concluded that "the motion of the earth alone, therefore, suffices to explain so many inequalities in the heavens."

Copernicus later described the "Order of the Spheres" in his heliocentric system, in which the time taken by a planetary sphere to make one revolution increases with the radius of its orbit:

The celestial spheres are arranged in the following order. The highest is the immovable sphere of the fixed stars, which contains and gives position to all things. Beneath it is Saturn, which Jupiter follows, then Mars. Below Mars is the sphere on which we revolve, then Venus; last is Mercury. The lunar sphere revolves around the center of the earth and moves with the earth like an epicycle. In the same order, also, one planet surpasses another in speed of revolution, accordingly as they trace greater or smaller circles. Thus Saturn completes its revolution in thirty years, Jupiter in twelve, Mars in two and one-half, and the earth in one year; Venus in nine months, Mercury in three.

Thus Mars, Jupiter, and Saturn revolve around the sun in orbits larger in size and longer in period than the earth's, while Mercury and Venus have smaller orbits with shorter periods. Copernicus shows how this can be used to explain the apparent retrograde motion of the planets, starting with Mars, Jupiter, and Saturn. He says each of the outer planets:

seems from time to time to retrograde, and often to become stationary. This happens by reason of the motion, not of the planet, but of the earth changing its position in the grand circle. For since the earth moves more rapidly than the planet, the line of sight directed [from the earth] toward [the planet and] the firmament regresses, and the earth more than neutralizes the motion of the planet. This regression is most notable when the earth is nearest to the planet, that is, when it comes between the plant and the sun at the evening rising of the planet. On the other hand, when the planet is setting in the evening or rising in the morning, the earth makes the observed motion greater than the actual. But when the line of sight is moving in the direction opposite to that of the planets and at an equal rate, the planets appear to be stationary, since the opposed motions neutralize each other.

The apparent retrograde motion of the inner planets, focusing on Venus, he explained as follows:

Venus seems at times retrograde, particularly when it is nearest the earth, like the superior planets, but for the opposite reason. For the regression of the superior planets happens because the motion of the earth is more rapid than theirs, because it [the earth] is slower; and because the superior planets enclose the grand circle [earth's orbit], whereas Venus is enclosed within it. Hence Venus is never in opposition to the sun [i.e., on the opposite side of the sun], and since the earth cannot come between them, but it [Venus] moves within fixed distances on either side of the sun. These distances are determined by tangents to the circumference drawn from the center of the earth, and never exceed 48° in our observations.

Copernicus used the same system of epicycles that Ptolemy and all of his successors had employed in the geocentric model. He did so because this was a highly effective method for representing the orbits of the planets as seen from the earth, which he adapted to his heliocentric system by changing the reference point to the sun as the center rather than the earth, which then became one of the planets. This was a stroke of genius, I think, for now he could put the mathematical methods of Ptolemy to use in his heliocentric theory.

Copernicus concluded the *Commentariolus* by summarizing the number of circles, that is, deferents, or primary circles, and epicycles, or secondary loops, required to describe all of the planetary motions in his heliocentric system: "Then Mercury runs on seven circles in all; Venus on five; the earth on three, and round it the moon on four; finally Mars, Jupiter and Saturn on five each. Altogether, therefore, thirty-four circles suffice to explain the entire structure of the universe and the entire ballet of the planets."

Despite its brevity, the *Commentariolus* contains all of the main ideas that Copernicus would later introduce in *De revolutionibus*, most notably the revelation that the sun and not the earth was the center of the universe. Although he says in the beginning of the *Commentariolus* that he will leave mathematical demonstrations for his larger work, he describes the motions of the celestial bodies at some length, indicating that he had already worked out the details of his system, though he would alter parts of his planetary model in the final version of *De revolutionibus*.

Thus some three decades before the appearance of *De revolutionibus* Copernicus had fully developed his heliocentric theory. This involved a tremendous amount of work, which he would have done in the first decade after his return from Italy, during most of which he was extremely busy in the service of his uncle Lucas. It would then seem as if he indeed left his uncle's service so as to return to the cathedral at Frauenburg, where he would have more time for his astronomical observations and calculations.

Aside from the mathematical details of the Ptolemaic and Copernican theories, which even modern astronomers find formidable, it seems to me that Copernicus conceived the truly revolutionary idea of the rotating sun at the center of the universe very early in his career, probably when he was at the University of Cracow. There he had become deeply influenced by the Pythagorean notion of celestial harmony, which he expressed throughout his work, and he was also aware of the heliocentric theory that Aristarchus of Samos had put forward in around 250 BC. Since he took the trouble to acquire and bind together the *Alphonsine Tables* with the works of Peurbach and Regiomontanus, we can be sure that Copernicus had followed the developing ideas of his medieval predecessors. So I am convinced that from his undergraduate days Copernicus was committed to the idea that the sun and not the earth was the center of the universe, a revolutionary theory that he spent the rest of his days developing mathematically and verifying by observation, applying the scientific method developed by Robert Grosseteste and his followers. Thus Copernicus represents the culmination of medieval European science and at the same time the beginning of the Scientific Revolution, working away quietly in his "obscure corner of the earth."

The first indication that the new theories of Copernicus had reached Rome came in the summer of 1533, when the papal secretary Johann Widmanstadt gave a lecture entitled *Copernicana de motuu terra sentential explicani* (*An Explanation of Copernicus's Opinion of the Earth's Motion*) before Pope Clement VI and a group that included two cardinals and a bishop. After the death of Pope Clement, on September 25, 1534, Widmanstadt entered the service of Cardinal Nicholas Schönberg, who as papal nuncio in Prussia and Poland had undoubtedly heard of Copernicus years before. Schönberg wrote to Copernicus on November 1, 1536, in a letter that may have been drafted by Widmanstadt, urging Copernicus to publish a book on his new cosmology and to send him a copy.

Despite this encouragement Copernicus made no move to publish his research. His attitude abruptly changed, though, in the spring of 1539, when he received an unexpected visit from a young German scholar, Georg Joachim van Lauchen, who called himself Rheticus (1514–1574). Rheticus, only twenty-five, already professor of mathematics at the Protestant University of Wittenberg, explained that he was deeply interested in the new cosmology of Copernicus, who received him hospitably and permitted him to study the manuscript that he had written to explain his theories. During the next ten weeks Rheticus studied the manuscript alongside Copernicus. He then summarized the work in a treatise entitled *Narratio prima (First Narrative)*, intended as an introduction to the Copernican theory. This was written in the form of a letter from Rheticus to his friend Johann Schöner, under whom he had studied in Wittenberg. The *Narratio prima* was published at Danzig in 1540 with the approval of Copernicus, who was referred to by Rheticus as "my teacher" in the introductory section where he described the scope of the Copernican cosmology.

Rheticus then went into each of the books in detail, adding an astrological prediction of his own after his account of the Copernican theory of "The Eccentricity of the Sun and the Motion of the Solar Apogee." Rheticus believed that world history followed the same cycle as the eccentricity of the sun's orbit (observed from the earth) and that the completion of its next cycle would coincide with the downfall of the Mohammedan faith, following which, he says, "We look forward to the coming of our Lord Jesus Christ when the center of the eccentric reaches the other boundary of mean value, for it was in that position at the creation of the world."

Rheticus does not mention the heliocentric theory until after the section on "General Considerations Regarding the Motions of the Moon, Together with the New Lunar Hypotheses." There he said that the new model explains the retrograde motion of the planets "by having the sun occupy the center of the universe, while the earth revolves instead of the sun on the eccentric."

The *Narratio prima* proved to be so popular that a second edition was published in Basel the following year. But Copernicus still hesitated to publish his manuscript, which he sent for safekeeping to his old friend Tiedemann Giese, bishop of Culm. A letter, now lost, that Copernicus wrote on July 1, 1540, indicates that he was afraid that his theory would be criticized by both Aristotelians and theologians, for it contradicted the accepted geocentric astronomical of Ptolemy as well as the world picture of both Catholic and Protestant theologians, and at a most critical time, at the beginning of the Reformation. At a more fundamental level, or so I believe, was Copernicus's fear of being ridiculed for proposing such a counterintuitive theory, particularly since he was such a private and modest person. As we will see, he admitted this in the dedicatory preface of *De revolitionibus*, addressed to Pope Paul III.

Finally, in the autumn of 1541, Giese received permission from Copernicus to send his manuscript to Rheticus, who took it to the press of Johannes Petreius in Nuremberg for publication. The title chosen for the book was *De revolutionibus orbium coelestium libri VI* (*Six Books Concerning the Revolutions of the Heavenly Spheres*).

Toward the end of the following year Copernicus suffered a series of strokes that left him half-paralyzed, and it was obvious to his friends that his end was near. Tiedemann Giese wrote on December 8, 1542, to George Donner, one of the canons at Frauenburg, asking him to look after Copernicus in his last illness. "I know that he always counted you among his truest friends. I pray therefore, that if his occasions require, you will stand by him and take care of the man whom you, with me, have ever loved, so that he may not lack brotherly help in his distress, and that we may not appear ungrateful to a friend who has richly deserved our love and gratitude."

Meanwhile Rheticus had taken a leave of absence from the University of Wittenberg in May 1542 to supervise the printing of *De revolutionibus* in Nuremberg. Five months later he left Nuremberg to take up a post at the University of Leipzig, leaving responsibility for the book in the hands of Andreas Osiander, a local Lutheran clergyman. Osiander took it upon himself to add an anonymous introduction entitled *Ad lectorem* (*To the Reader*), which was to be the cause of considerable controversy regarding the Copernican theory.

De revolutionibus finally came off the press in the spring of 1543, with the publisher's blurb, probably also written by Osiander, printed directly below the title. "You have in this recent work, studious reader, the motion of both the fixed stars and the planets recovered from ancient as well as recent observations and outfitted with wonderful new and admirable hypotheses. You also have most expeditious tables from which you can easily compute the positions of the planets for any time. Therefore buy, read, profit."

The first printed copy of *De revolutionibus* was sent to Copernicus, and according to tradition it reached him a few hours before he died, on May 24, 1543. Tiedemann Giese described the last days of Copernicus in a letter to Rheticus: "He had lost his memory and mental vigor many days before; and he saw his completed work only at his last breath upon the day that he died."

The introduction to *De revolutionibus,* the *Ad lectorum* written by Osiander, is entitled "To the Reader Concerning the Hypotheses of This Work." Here, Osiander cautions that the book is designed as a mathematical device for calculation and not as a real description of nature. The *Ad lectorum* was intended to deflect criticism of the heliocentric cosmology by those who thought that it contradicted the Bible, particularly the passage in the book of Joshua that says, "The sun stood still in the middle of the sky and delayed its setting for almost a whole day." Martin Luther, referring to the Copernican theory, had already been quoted as saying that "people give ear to an upstart astrologer who strove to show that the earth revolves, not the heavens, or the firmament, the sun and the moon. This fool wishes to reverse the entire science of astronomy, but sacred Scripture tells us that Joshua commanded the Sun to stand still and not the Earth." Copernicus himself had been worried about such criticism, as evidenced by his statement in the preface of *De revolutionibus,* which he dedicated to Pope Paul III: "I can reckon easily enough, Most Holy Father, that as soon as certain people learn that in these books of mine which I have written about the revolutions of the spheres of the world I attribute certain motions to the terrestrial globe they will immediately shout to have me and my opinion hooted off the stage."

De revolutionibus begins with a greatly simplified description of the Copernican cosmology and its philosophical basis. Copernicus shows how the rotation of the earth on its axis, together with its revolution about the sun, can easily explain the observed motions of the celestial bodies. The absence of stellar parallax he explains by the fact that the radius of the earth's orbit is negligible compared to the distance of the fixed stars, an argument first used by Archimedes in discussing the heliocentric theory proposed by Aristarchus. All of the arguments on physical grounds against the earth's motion are then refuted, using in most cases the explanations given by Nicholas of Cusa. Copernicus abandons the Aristotelian doctrine that the earth is the sole source of gravity and instead takes the first step toward the Newtonian theory of universal gravitation:

> I myself think that gravity or heaviness is nothing except a certain natural appetency implanted in the parts by the divine providence of the universal Artisan, in order that they should unite with one another in their oneness and wholeness and come together in the form of a globe. It is believable that this affect is present in the sun, moon, and the other bright planets and that through its efficacy they remain in a spherical figure in which they are visible, though they nevertheless follow their circular motions in many different ways.

One of the advances made by Copernicus deals with the ambiguity concerning Mercury and Venus, which in the Ptolemaic model were sometimes placed "above" the sun and sometimes "below." The Copernican system has Mercury as the closest planet to the sun, followed by Venus, Earth, Mars, Jupiter, and Saturn, surrounded by the sphere of the fixed stars, and with the moon orbiting the earth. This model is simpler and more harmonious than Ptolemy's, for all of the planets revolve in the same sense, with velocities decreasing with their distance from the sun, which sits enthroned at the center of the cosmos:

In the center of all the celestial bodies rests the Sun. For who would place this lamp of a very beautiful temple in another or better place than this wherefrom it can illuminate everything at the same time. As a matter of fact, not unhappily do some call it the lantern, others the mind and still others, the pilot of the world.... And so the sun, as if resting on a kingly throne, governs the family of stars which wheel around.

Copernicus refers to what he calls "the three motions" of the earth. These are the earth's rotation on its axis, its revolution around the sun, and a third conical motion, which he introduced to keep the earth's axis pointing in the same direction while the crystalline sphere in which it was embedded rotated annually. The period of this supposed third motion he took to be slightly different than the time it takes the earth to rotate around the sun, the difference being due to the very slow precession of the equinoxes.

Copernicus goes on to give a detailed introduction to astronomy and spherical trigonometry, together with mathematical tables and a catalogue of the celestial coordinates of 1,024 stars, most of them derived from Ptolemy, adjusted for the precession of the equinoxes. This is the data to which he would apply his mathematical theory, which was much the same as that of Ptolemy, but with the sun at the center rather than the earth, which thus becomes one of the planets.

Copernicus then turned his attention to the precession of the equinoxes and the movement of the earth around the sun. Here the theory is unnecessarily complicated, since Copernicus, besides combining precession with his "third motion" of the earth, inherited two effects from his predecessors, one of them spurious. The first effect was the mistaken notion, stemming from the trepidation theory, that the precession was not constant but variable, and the other was the variation in the inclination of the ecliptic. As a result, his explanation of precession is flawed.

Another problem that Copernicus faced was the irregularity in the sun's motion as observed from the earth. This is due to the fact that the center of the orbit does not coincide with the center of the sun and that the velocity of the sun in its apparent orbit around the earth is variable. He solved the problem by using a combination of an eccentric circle and an epicycle, in this way avoiding the use of an equant, his principal objection to Ptolemy's mathematical method.

Next Copernicus dealt with the irregularity in the motion of the moon around the earth. Here he solved the problem by using an eccentric circle along with an epicycle together with a second, smaller epicyclet carrying the moon, similar to the model that had been used by the Arabic astronomer ibn al-Shatir, which Copernicus may have seen when he was studying at the University of Cracow.

He then went on to study the motions of the planets. Here again Copernicus used eccentrics and epicycles just as Ptolemy had done, though his conviction that the celestial motions were combinations of circular motion at constant angular velocity made him refrain from using the Ptolemaic device of the equant. Because of the complexity of the celestial motions, Copernicus was forced to use about as many circles as had Ptolemy, and so there was little to choose from between the two theories so far as economy was concerned, and both were capable of giving results of comparable accuracy. The advantages of the Copernican system were that it was more harmonious, it removed the ambiguity about the order of the inner planets, it explained the retrograde motion of the planets as well as their variation in brightness, and it allowed both the order and sizes of the planetary orbits to be determined from observation without any additional assumptions.

Historian Angus Armitage noted that the great contribution that Copernicus made to astronomy was not just the notion that the sun was the center of the cosmos, but "lay rather in his development of those ideas into a systematic theory, capable of furnishing tables of an accuracy not before attained, and embodying a principal the adoption of which was to make possible the triumphs of Kepler and Newton in the following century."

The sizes of the planetary orbits were determined relative to the mean radius of the earth's orbit, known today as an astronomical unit (a.u.), where for each planet he computed the greatest, mean, and least values.

The distance from the sun of either Mercury and Venus, the inner planets, was determined from its maximum elongation from the sun, that is, the largest angle it makes with the sun as seen from the earth. In the case of Venus the maximum elongation is 48°, from which Copernicus computed that its mean distance from the sun was 0.719 a.u., which is in remarkably good agreement with the modern value of 0.723 a.u. He used the same method for Mercury, where the maximum elongation is 22°, determining that its mean distance from the sun was 0.376 a.u., as compared to the modern value of 0.387 a.u., the larger discrepancy being due to the greater difficulty in observing Mercury because it is so close to the sun.

Copernicus determined the orbital radii of the outer planets using slightly different geometrical procedures. The values that he found for the mean values, in astronomical units, were very close to the modern values, an extraordinary accomplishment considering the primitive instruments he was using.

The next task was to convert these relative distances to absolute values. Using the methods of Aristarchus, Ptolemy, and al-Battani, along with his own innovations, Copernicus found that the sun's mean distance was 1,142 earth radii (e.r.), as compared to Ptolemy's value of 1,179 e.r. He could then have converted the known radius of the earth into miles, but he chose not to do so, for he felt that his new planetary model was so complete and self-consistent that such tasks could be left to his followers. In his dedicatory preface to Pope Paul III he stressed the internal consistency of his theory.

> And so, having laid down the movements which I attribute to the Earth farther on in the work, I finally discovered by the help of long and numerous observations that if the movements of the other wandering stars are correlated with the circular movement of the Earth, and if the movements are computed in accordance with the revolution of each planet, not only do all their phenomena follow from that but also this correlation binds together the order and magnitude of all the planets and of their spheres or orbital circles and the heavens themselves that nothing can be shifted around in

any part of them without disrupting the remaining parts and the universe as a whole.

Copernicus mentions some of the Arabic astronomers whose observations and theories he used in *De revolutionibus*, namely al-Battani, al-Bitruji, al-Zarqali, ibn Rushd (Averroës), and Thabit ibn-Qurra. He does not refer to ibn al-Shatir, though in his lunar theory he used a model similar to one that his predecessor had developed. Neither does he cite the thirteenth-century Arabic astronomer Nasir al-Din al-Tusi, although recent research shows that Copernicus used a mathematical method that had been invented by him. This is the so-called al-Tusi couple, which Copernicus also used in his lunar theory. There is no definite evidence that Copernicus knew of al-Tusi's theory, which was apparently known to some of his contemporaries.

Copernicus referred to Aristarchus of Samos six times in the original manuscript of *De revolutionibus*, but four of these were erroneous and a fifth was probably incorrect. The only correct reference was near the end of book 1, where Copernicus wrote:

Though the courses of the Sun and the Moon can surely be demonstrated on the assumption that the Earth does not move, it does not work so well with the other planets. Probably for this and other reasons, Philolaus perceived the mobility of the Earth, a view also shared by Aristarchus of Samos, so some say, not impressed by that reasoning which Aristotle cites and refutes. Yet, since only keen wits and long efforts can probe such things, it was then hidden from most philosophers, and, as Plato said, only a few grasped the real cause of planetary motion.

But in the final editing of the manuscript this passage was removed and, instead, another passage was rewritten into the preface to Pope Paul III, in which Copernicus mentions Greek astronomers who had the earth in motion, though omitting Aristarchus. The new passage reads:

> Some think that the Earth is at rest; but Philolaus the Pythagorean
> says that it moves around the fire, with an obliquely circular motion,
> like the sun and moon. Herakleides of Pontus and Ekphantus the
> Pythagorean do not give the earth any movement of locomotion,
> but rather a limited movement of rising and setting around its cen-
> ter like a wheel.

Copernicus cited Plutarch as the reference for this passage, but the source is actually from a work by Aetius (Pseudo-Plutarch) entitled *Placita philosophorum*. Copernicus is known to have possessed a copy of George Valla's *Outline of Knowledge*, printed by Aldus Manutius in Venice in 1501, which included the *Placita*. The *Placita* contained two other references concerning the motion of the earth, neither of which was used by Copernicus. One has Aristarchus "assuming that the heavens are at rest while the earth revolves along the ecliptic, simultaneously rotating about its own axis"; the other says that in his theory the earth "spins and turns, which Seleucus afterwards advanced as an established opinion."

Very little was known in western Europe of the heliocentric theory of Aristarchus until the publication of Archimedes' *Sand Reckoner* in 1544, the year after Copernicus died. According to astronomer Owen Gingerich: "Had Copernicus known more [about Aristarchus' heliocentric theory] he surely would have been happy to mention it, since he needed all the support he could muster for his own unorthodox views, and since he quotes with enthusiasm other possible geokineticists from Antiquity with less reputable credentials."

Gingerich went on to say: "There is no question but that Aristarchus had the priority of the heliocentric idea. Yet there is no evidence that Copernicus owed him anything. As far as we can tell, both the idea and its justification were found independently by Copernicus."

There is no question in my own mind that Copernicus conceived his sun-centered cosmology on his own, and that had he known more about Aristarchus he certainly would have given him credit. So far as I am concerned, Gingerich says the last word on the question of who should get credit for the heliocentric theory:

> It is not really the task of the historian of science to assess the comparative originality of these two scientific giants. The heliocentric cosmology was convincing neither to the contemporaries of Aristarchus nor to those of Copernicus, but Copernicus had the good luck to be born not only at a time when science was beginning to reach, so to speak, a critical mass, but also at a time when scientific works were beginning to be printed; therefore his arguments survived and convinced a later generation of astronomers.

Astronomy and cosmology would never again be the same after the publication of *De revolutionibus*. The world picture was now irrevocably changed, an intellectual revolution started by an obscure canon, reviving a theory that had first been proposed eighteen centuries before by an almost forgotten Greek astronomer.

But Copernicus would not be forgotten because the revolution that he started made him immortal. He was certainly my hero when I first began reading books on astronomy in my youth. My admiration was and still is not just for his scientific achievement but for his persistent courage in working away on his revolutionary theory in his "remote corner of the earth," his life's work rescued from oblivion in the last years of his life by an idealistic young mathematician who saw to the printing of *De revolutionibus*, planting the seed that would blossom in the new astronomy and physics of the following century.

Copernicus never taught at a university, and so he never had a pupil in the formal sense, but Rheticus was in effect his student when they worked together on the final version of *De revolutionibus*, and he is so listed in the Math-Physics Genealogy website. And it is through this tenuous link that I and my students and their students trace our scientific ancestry back to Copernicus through Galileo and Newton.

12

The New Astronomy

*T*HE CLIMATE OF OPINION IN REFORMATION EUROPE, BOTH RELI-gious and intellectual, was unfavorable for the acceptance of the Copernican heliocentric theory, which was opposed by both the Catholic and the Protestant churches as well as scholastic philosophers in the universities, who held firmly to the geocentric cosmology of Aristotle and Ptolemy.

The gradual acceptance of the Copernican theory was due primarily to its success in giving an accurate mathematical description of the motion of the celestial bodies. His sun-centered cosmology, by contrast, was in most cases either ignored or rejected outright. The gradual acceptance of his theories was due to their use in new astronomical tables that were more successful in predicting celestial motion than the old ones based on the Ptolemaic theory.

The first astronomical tables based on the Copernican theory were courtesy of Erasmus Reinhold (1511–1553), professor of astronomy at the University of Wittenberg at the same time as Rheticus. When Rheticus returned to Wittenberg in September 1541 with the manuscript of *De revolutionibus*, he likely showed it to Reinhold. The following year Reinhold published his commentary on Peurbach's *Theoricae novae planetarum*, in which he wrote of Copernicus in the preface: "I know of a modern scientist who is exceptionally skillful. He has raised a lively expectancy in everybody. One hopes that he will restore astronomy." Later he wrote, "I hope that this astronomer, whose genius all posterity will rightly admire, will at long last come to us from Prussia."

Reinhold set out to produce a more extensive version of the planetary tables in *De revolutionibus*, which were published in 1551 as the *Prutenic Tables*, where in the introduction he praises Copernicus but is silent about his heliocentric theory. Reinhold also wrote a commentary on *De revolutionbus*, but it was unfinished when he died of the plague in 1553.

The *Prutenic Tables* were the first complete planetary tables prepared in Europe for three centuries. They were demonstrably superior to the older tables, which were now out of date, and so they were used by most astronomers, lending legitimacy to the Copernican theory even when those who used them did not acknowledge the sun-centered cosmology of Copernicus. As the English astronomer Thomas Blundeville wrote in the preface to an astronomy text in 1594: "Copernicus ... affirmeth that the earth turneth about and that the sun standeth still in the midst of the heavens, by help of which false supposition he hath made truer demonstrations of the motions and revolutions of the celestial spheres, than ever were made before."

The English mathematician Robert Recorde (1510–1558) was one of the first to lend some support to the Copernican theory. He discusses the theory in his *Castle of Knowledge* (1551), an elementary text on Ptolemaic astronomy with a brief favorable reference to the Copernican theory. It is written in the form of a dialogue between a master and a scholar concerning Ptolemy's arguments against the earth's motion. After the scholar sums up these arguments, the master presents the Copernican theory in a very positive manner.

> That is trulye to be gathered: howe be it, COPERNICUS, a man of greate learning, of much experience, and of wonderfull diligence in observation, hath renewed the opinion of ARISTARCHUS SAMIUS, and affirmith that the earthe not only moveth circular-lye about his own centre, but also may be, yea and is, continually out of the precise centre of the world 38 hundred thousand miles.

Five years later an *Ephemeris*, or set of astronomical tables, was printed in London for the year 1557 by John Feild, based on the work of Copernicus and Reinhold. The introduction was written by John Dee (1527–1608), Queen Elizabeth's astrologer, who noted that he had persuaded Feild to compile the *Ephemeris*, based on the Copernican theory, since the older tables were no longer satisfactory. Dee praised Copernicus for his "more than Herculean" effort in restoring astronomy, though he said that this was not the place to discuss the heliocentric theory itself.

A second edition of *De revolutionibus* was published in Basel in 1566, and copies of it made their way to Italy and England. The English astronomer Thomas Digges (c. 1546–1595), a pupil of Dee, obtained a copy of *De revolutionibus* that has survived in the library of Geneva University, along with a note he wrote on the title page, *"Vulgi opinio Error"* ("the common opinion errs"), indicating that he was one of the few sixteenth-century scholars who accepted the Copernican theory.

Digges had already published a work on the nova or new star of 1572. (It is now known that a nova is a star that explodes at the end of its evolutionary cycle, releasing an enormous amount of energy for a few months.) He hoped to use his observations to confirm or refute the Copernican theory, and he appealed to other astronomers to help him.

Digges did a free English translation of chapters 9 through 11 of the first book of *De revolutionibus*, adding it to his father's perpetual almanac, *A Prognostication Everlasting*, publishing them together in 1576 as a *Perfit Description of the Caelestiall Orbes, according to the most ancient doctrines of the Pythagoreans lately revived by Copernicus and by Geometricall Demonstrations approved*. Digges stated that he had included this excerpt from *De revolutionibus* in the almanac "so that Englishmen might not be deprived of so noble a theory."

The book was accompanied by a large folded map of the sun-centered universe, in which the stars were not confined to the outermost celestial sphere but scattered outward indefinitely in all directions. Digges thus burst the bounds of the medieval cosmos, which till then had been limited by the ninth celestial sphere, the one containing the supposedly fixed stars, which in his model extended to infinity.

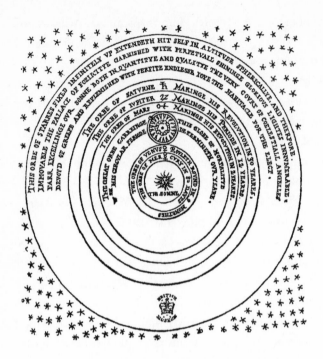

A model of the Copernican system from Thomas Digges's
A Prognostication Everlasting, *1576*

The geocentric celestial spheres were still very much a part of the general worldview in late-sixteenth-century England, as is evident from Christopher Marlowe's play, *The Tragical History of Dr. Faustus*. As soon as Faustus has made his pact with the devil, he begins to question Mephistopheles about matters beyond the ken of mortals, beginning with the celestial spheres.

> Speak, are there many spheres above the moon?
> Are all celestial bodies but one globe,
> As is the substance of this centric earth?

The concept of an infinite universe was one of the revolutionary ideas for which the Italian mystic Giordano Bruno (1548–1600) was condemned by the Catholic Church, which had him burned at the stake in Rome on February 17, 1600. Bruno expressed his belief in the Copernican theory on several occasions, most notably in a lecture he gave at Oxford in 1583. Historian Frances A. Yates describes Bruno's mystical interpretation of the theory, in which he was encouraged by the reference that Copernicus had made to Hermes Trismegistus:

> In his defense of Copernicanism against the Aristotelians of Oxford, Bruno presented Copernicus "only as a mathematician" who had not appreciated the true inwardness of his discovery as he, Bruno, understood it—as portending a return to magical insight into living nature. In support of the movement of the earth, Bruno quoted a passage from one of the treatises of the *Corpus Hermeticum*, which states that the earth moves because it is alive.

Yates went on to say that "the legend that the nineteenth century built around Bruno as the hero who, unlike Galileo, refused to retract his belief that the earth moves is entirely without foundation. Bruno's case, however, may have affected the attitude of the Church toward the Copernican hypothesis, and may have encouraged the Inquisition's suspicion of Galileo."

At the beginning of his dialogue on *The Infinite Universe and the Worlds*, published in 1584, Bruno said, through one of his characters, that in this limitless space there are innumerable worlds similar to our earth, each of them revolving around its own star-sun. "There are then innumerable suns, and an infinite number of earths revolve around these suns, just as the seven [the five visible planets plus the earth and its moon] we can observe revolve around this sun which is close to us."

Bruno's universe was not only infinite but dynamic, in contrast to the finite cosmos of Aristotle, for whom the celestial region was immutable. Here Bruno took his inspiration from the atomic theory of Democritus and Epicurus as interpreted by Lucretius in his *De rerum natura*, which had been rediscovered in 1417. According to Bruno, the living universe is limited neither in its extent nor in its constantly changing multiplicity:

There are no ends, boundaries, limits or walls which can defraud or deprive us of the infinite multitude of things.... Thus Democritus and Epicurus, who maintained that everything throughout infinity suffered renewal and restoration, understood these matters more truly than those who at all costs maintain a belief in the immutability of the Universe, alleging a constant and unchanging number of particles of identical material that perpetually undergo transformation, one into another.

The concept of an infinite universe appears also in the work of the English scientist William Gilbert (1544–1603), who may have been influenced in this regard by Thomas Digges and Giordano Bruno. Gilbert's *De magnete*, published in 1600, was the first work on magnetism since that of Petrus Peregrinus in the thirteenth century.

The sixth and final book of *De magnete* was devoted to Gilbert's cosmological theories, in which he rejected the crystalline celestial spheres of Aristotle and said that the apparent diurnal rotation of the stars was actually due to the axial rotation of the earth, which he believed to be a huge magnet. His rejection of diurnal stellar motion was based on his belief that the stars were limitless in number and extended to infinity, so that it was thus ridiculous to assume that they rotated nightly around the celestial pole. While Gilbert discussed the motions of the earth according to Copernicus and Giordano Bruno, he did not affirm or deny the heliocentric theory; at times he dismissed it as not being pertinent to the topic he was discussing.

Meanwhile astronomy was being revolutionized by the Danish astronomer Tycho Brahe (1546–1601), who in the last quarter of the seventeenth century made systematic observations of significantly greater accuracy than any ever done in the past, all just before the invention of the telescope.

Tycho, born to a noble Danish family, was brought up by his paternal uncle, Jörgen Brahe, who hired a tutor to teach him Latin and the preparatory subjects for a university education. He enrolled at the University of Copenhagen at the age of thirteen and subsequently continued his studies at the universities of Leipzig, Rostock, and Basel, from which he graduated in 1568. The astronomy books that he studied included Sacrobosco's *De sphaera*, which had been in use since the thirteenth century, and other texts based on the homocentric spheres of Aristotle and the epicycles and eccentrics of Ptolemy.

There had been an eclipse of the moon on October 28, 1566, that led Tycho to predict the death of the Ottoman sultan Süleyman the Magnificent, who, as they soon learned, had already died on September 6 of that year. On December 29 Tycho's third cousin, Manderup Parsburg, mocked him about this at a dinner party and they both drew their swords and went outside to duel. Before the other guests could stop them, Parsburg had lopped off a large part of Tycho's nose, which, after a long convalescence, he replaced with an artificial nose of gold and silver blended to appear flesh colored. The disfigurement is apparent in the portrait of Tycho that appears in his book *Astronomiae instauratae mechanica*, published in 1598.

The occurrence of a solar eclipse on August 21, 1560, though only partial in Copenhagen, aroused Tycho's interest in observational astronomy, which was not part of the university curriculum. He immediately obtained a copy of a set of *ephemerides*, or planetary tables, based on Reinhold's *Prutenic Tables*. Here he first became aware of the work of Copernicus, whom he called "a second Ptolemy." Tycho made an observation in Leipzig in August 1563 that he considered to be a turning point in his career, when he noted a conjunction of Saturn and Jupiter, measuring the angular distance between the two planets using a pair of compasses. He found that the *Alphonsine Tables* were a month off in predicting the date of the conjunction, and that the *Prutenic Tables* were several days in error. He began making observations with a homemade instrument known as a radius or cross-staff, and since it was not very accurate, he devised a table of corrections to make up for its deficiencies.

Early in 1569 Tycho went to Augsburg, in Germany, where he made his first observation on April 14. The instruments that he designed and built for his observations included a great quadrant with a radius of some nineteen feet for measuring the altitude of celestial bodies. He also constructed a huge sextant with a radius of fourteen feet for measuring angular separations, as well as a celestial globe ten feet in diameter on which to mark the positions of the stars in the celestial map that he began to create.

Tycho returned to Denmark in 1570 and, after his father's death on May 9 of the following year, he moved to Herrevad Abbey, the home of his maternal uncle, Steen Bille, where he spent his time performing alchemical experiments. The following year he entered a common-law marriage with Kirsten Jorgensdatter, the daughter of the local pastor, who was to bear him six children.

On November 11, 1572, as Tycho was walking back to supper from his laboratory, he saw a nova, or new star, that had suddenly appeared in the constellation Cassiopeia, exceeding even the planet Venus in its brilliance. As he later wrote in his tract on the nova: "I knew perfectly well— for from my earliest youth I have known all the stars in the sky, something which one can learn without difficulty—that no star had ever existed in that place in the heaven, not even the very tiniest, to say nothing of a star of such striking clarity."

Tycho's measurements indicated that the nova was well beyond the sphere of Saturn, and the fact that its position did not change showed that it was not a comet. This was clear evidence of a change taking place in the celestial region, where, according to Aristotle's doctrine, everything was perfect and immutable.

The nova eventually began to fade, its color changing from white to yellow and then red, finally disappearing from view in March 1574. By then Tycho had written a brief tract entitled *De nova stella* (*The New Star*), which was published in Copenhagen in May 1573. After presenting the measurements that led him to conclude that the new star was in the heavens beyond the planetary spheres, Tycho expressed his amazement at what he had observed. "I doubted no longer," he wrote. "In truth, it was the greatest wonder that has ever shown itself in the whole of nature since the beginning of the world, or in any case as great as [when the] Sun was stopped by Joshua's prayers."

The tract impressed King Frederick II of Denmark, who gave Tycho an annuity along with the small offshore island of Hveen, in the Oresund Strait north of Copenhagen, the revenues of which would enable him to build and equip an observatory. Tycho settled on Hveen in 1576, calling the observatory Uraniborg, meaning "City of the Heavens." The astronomical instruments and other equipment of what came to be a large research center were so numerous that Tycho was forced to build an annex called Stjernborg, "City of the Stars," with subterranean chambers to shield the apparatus and researchers from the elements. That same year Tycho and his assistants began a series of observations of unprecedented accuracy and precision that would continue for the next two decades, laying the foundations for what would prove to be the new astronomy.

Tycho's main project at Uraniborg was to make new and more accurate determinations of the celestial coordinates of the fixed stars, and to observe the changing positions of the sun, moon, and planets for the purpose of improving the theories of their motions. He produced a catalogue giving the coordinates of 777 fixed stars, to which he later added another 223 so as to bring the total to 1,000.

The celestial coordinates given in Tycho's star catalogue had a mean error, compared to modern values, of less than 40 seconds of arc, far less than that of any of his predecessors. Comparing the coordinates of the twenty-one principal stars in his catalogue with those measured from antiquity up to his own time, Tycho computed a value for the rate of precession of the equinoxes equal to 51 seconds of arc per year, as compared to the modern value of 50.23 seconds. He correctly assumed the precession to be uniform, making no mention of the erroneous Islamic trepidation theory that had caused unnecessary problems for Copernicus.

Shortly after sunset on November 13, 1577, Tycho first noticed a spectacular comet with a very long tail, and he continued to observe it nightly until January 26 of the following year, by which time it had faded to the point that it was hardy visible. His detailed observations of the comet led him to conclude that at its closest approach it had been farther away than the moon, in fact even beyond the orbit of Mercury, and that it was in an oval orbit around the sun among the outer planets. This contradicted the Aristotelian doctrine that comets were meteorological phenomena occurring below the sphere of the moon. Tycho was thus led to reject Aristotle's concept of the homocentric crystalline spheres, and he concluded that the planets were moving independently through space.

> There really are not any spheres in the heavens.... Those which have been devised by the experts to save the appearances exist only in the imagination for the purpose of enabling the mind to conceive the motion which the heavenly bodies trace in their course and, by the aid of geometry, to determine the motion numerically through the use of arithmetic.

Despite his admiration for Copernicus, Tycho rejected the heliocentric theory, both on physical grounds and on the absence of stellar parallax, where in the latter case he did not take into account the argument made by Archimedes and Copernicus that the stars were too far away to show any parallactic shift. Tycho rejected both the diurnal rotation of the earth and its annual orbital motion, retaining the Aristotelian belief that the stars rotated nightly around the celestial pole.

Faced with the growing debate between the Copernican and Ptolemaic theories, Tycho proposed his own planetary model, which combines elements of the geocentric and heliocentric theories. In the Tychonic system the immobile earth is still at the center of the universe, with the sphere of the fixed stars revolving around it in twenty-four hours. The other planets were in orbit around the sun, which circled the earth every twenty-four hours, while the moon went around the earth once a month. The orbits of Mercury and Venus intersected the orbit of the sun at two points, but did not encompass the earth. The orbit of Mars also intersected the sun's orbit in two places, but enclosed the earth and its orbiting moon. The orbits of Jupiter and Saturn enclosed the entire path of the sun. Tycho believed that his model combined the best features of both the Ptolemaic and Copernican theories, since it kept the earth stationary and explained why Mercury and Venus were never very far from the sun.

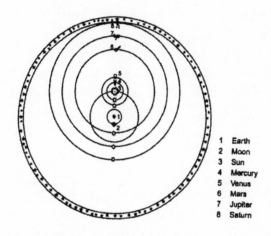

1	Earth
2	Moon
3	Sun
4	Mercury
5	Venus
6	Mars
7	Jupiter
8	Saturn

The Tychonic model for planetary motion

Tycho's patron Frederick II died in 1588 and was succeeded by his son Christian IV, who was then eleven years old. When Christian came of age, in 1596, he informed Tycho that he would no longer support his astronomical research. Without financial backing, Tycho was forced to abandon Uraniborg the following year, taking with him all of his astronomical instruments and records, hoping to find a new royal patron.

Tycho moved first to Copenhagen and then in turn to Rostock and Wandsburg Castle, outside Hamburg. He remained for two years at Wandsburg Castle, where in 1598 he published his *Astronomiae instauratae mechanica*, a description of all his astronomical instruments. He sent copies of his treatise to all of the wealthy and powerful people who might be interested in supporting his further researches. He appended his star catalogue to the copy he presented to the emperor Rudolph II, who agreed to support Tycho's work, appointing him as the court astronomer.

Thus in 1600 Tycho moved to Prague, where he set up his instruments and created a new observatory at Benatky Castle, several miles northeast of the city. Soon afterward he hired an assistant named Johannes Kepler (1571–1630), a young German mathematician who had sent him an interesting treatise on astronomy, the *Mysterium cosmographicum*.

Kepler was born on December 27, 1571, in Weil der Stadt in southwestern Germany. His father was an itinerant mercenary soldier, his mother a fortune-teller who at one point was accused of being a witch and almost burned at the stake. The family moved to the nearby town of Lemberg, where Kepler was enrolled in one of the excellent Latin schools founded by the Duke of Württemberg. Kepler's youthful interest in astronomy had been stimulated by seeing the comet of 1577 and a lunar eclipse in 1580.

In 1589 Kepler entered the University of Tübingen, where, in addition to his studies in mathematics, physics, and astronomy, he was influenced by Platonism, Pythagoreanism, and the cosmological ideas of Nicholas of Cusa. His mathematics lectures were based on the works of Euclid, Archimedes, and Apollonius of Perge. (As Kepler later said, "How many mathematicians are there, who would toil through the *Conics* of Apollonius of Perge?")

Kepler was particularly influenced by his professor of astronomy, Michael Maestlin, from whom he first learned of the heliocentric theory. In the introduction to his first book, the *Mysterium cosmographicum*, Kepler wrote of his excitement on discovering the work of Copernicus, which he described as "a still unexhausted treasure of truly divine insight into the magnificent order of the whole world and of all bodies." He wrote also of how Copernicus inspired him to begin thinking about the causes behind this cosmic order:

When I was studying under the distinguished Michael Maestlin at Tübingen six years ago, seeing the inconveniences of the commonly accepted theory of the universe, I became so delighted with Copernicus, whom Maestlin often mentioned in his lectures, that I often defended his opinions in the students' debates about physics. I even wrote a painstaking disputation about the first motion, maintaining that it happens because of the rotation of the earth. I have by degrees—partly out of hearing Maestlin, partly by myself—collected all the advantages that Copernicus has over Ptolemy. At last in the year 1595 in Graz when I had an intermission in my lectures, I pondered on the subject with the whole energy of my mind. And there were three things above all for which I sought the causes as to why it was this way and not another—the number, the dimensions, and the motion of the orbs.

Kepler received his master's degree in Tübingen in 1591, after which he studied theology there until 1594, when he was appointed teacher of mathematics at the Protestant seminary in the Austrian town of Graz. A year after his arrival in Graz, Kepler developed an idea that he thought explained the arrangement and order of the heliocentric planetary system. He had learned from his reading of Euclid that there were five and only five regular polyhedra, the so-called Platonic solids, in which all of the faces are equal as well as equilateral—the cube, tetrahedron, dodecahedron, icosahedron, and octahedron—and it occurred to him that they were related to the orbits of the earth and the five other planets. He explained the scheme in his treatise, the *Mysterium cosmographicum*, published in 1596:

> The earth's orbit is the measure of all things; circumscribe around it a dodecahedron, and the circle containing it will be Mars; circumscribe around Mars a tetrahedron, and the circle containing this will be Jupiter; circumscribe around Jupiter a cube, and the circle containing this will be Saturn. Now inscribe within the earth an icosahedron, and the circle contained in it will be Venus; inscribe

within Venus an octahedron, and the circle contained in it will be Mercury. You now have the reason for the number of planets.

The values that Kepler calculated for the relative radii of the planetary orbits agree fairly well with those determined by Copernicus, when allowances are made for the eccentricities of the orbits. Also, in the case of Mercury, Kepler compromised by taking the radius of a sphere formed by the edges of the octahedron rather than in the octahedron itself, though there was no physical basis for his theory. With these concessions everything agrees within 5 percent, except for Jupiter, where Kepler's value is off by 9 percent, at which, as he writes, "no one will wonder, considering such a great distance."

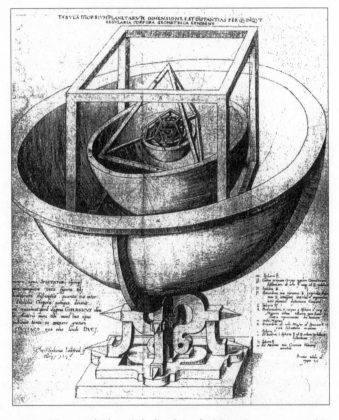

Drawing of the nested Platonic bodies from the Mysterium cosmographicum

Kepler sent copies of his treatise to a number of scientists, including Galileo Galilei (1564–1642). In his letter of acknowledgment, dated August 4, 1597, Galileo congratulated Kepler for having had the courage, which he himself lacked, of publishing a work supporting the Copernican theory. Kepler wrote back to Galileo on October 13, 1597, encouraging him to continue supporting the Copernican theory. "Have faith, Galilei, and come forward!" he wrote. "If my guess is right, there are but few of the prominent mathematicians of Europe who would wish to secede from us: such is the power of truth."

Kepler had also sent a copy of the *Mysterium cosmographicum* to Tycho Brahe, who received it after he had left Denmark for Germany. Tycho responded warmly, calling the treatise "a brilliant speculation," beginning a correspondence that eventually led Kepler to accept Tycho's invitation to join him at his new observatory at Benatky. As Tycho wrote in response to Kepler's letter of acceptance: "You will come not so much as a guest but as a very welcome friend and highly desirable participant and companion in our observations of the heavens."

Kepler finally arrived at Benatky with his family early in 1600, beginning a brief but extraordinarily fruitful collaboration with Tycho. When Kepler began work at Benatky he had hopes that he could take Tycho's data and use it directly to check his own planetary theory. But he was disappointed to find that most of Tycho's data was still in the form of raw observations, which first had to be subjected to mathematical analysis. Moreover, Tycho was extremely possessive of his data and would not reveal any more of it than Kepler needed for his work.

These and other disagreements with Tycho led Kepler to leave Benatky in April of that year, though he returned in October after considerable negotiation concerning the terms of his employment. Tycho then assigned Kepler the task of analyzing the orbit of Mars, which up to that time had been the responsibility of his assistant, Longomontanus, who had just resigned. Kepler later wrote: "I consider it a divine decree that I came at exactly the time when Longomontanus was busy with Mars. Because assuredly either through it we arrive at the knowledge of the secrets of astronomy or else they remain forever concealed from us."

Mars and Mercury are the only visible planets with eccentricities large enough to make their orbits significantly different from perfect circles. But Mercury is so close to the sun that it is difficult to observe, leaving Mars as the ideal planet for checking a mathematical theory, which is why Kepler was so enthusiastic at being able to analyze its orbit.

Early in the autumn of 1601 Tycho brought Kepler to the imperial court and introduced him to the emperor Rudolph. Tycho then proposed to the emperor that he and Kepler compile a new set of astronomical tables. With the emperor's permission, this would be named the *Rudolphine Tables*, and since it was to be based on Tycho's observations it would be more accurate than any done in the past. The emperor graciously consented and agreed to pay Kepler's salary in this endeavor.

Soon afterward Tycho fell ill, and after suffering in agony for eleven days, he died on October 24, 1601. On his deathbed he made Kepler promise that the *Rudolphine Tables* would be completed, and he expressed his hopes that it would be based on the Tychonic planetary model. As Kepler later wrote of Tycho's final conversation with him: "Although he knew I was of the Copernican persuasion, he asked me to present all my demonstrations in conformity with his hypothesis."

Two days after Tycho's death the emperor Rudolph appointed Kepler as court mathematician and head of the observatory at Benatky. Kepler thereupon resumed his work on Mars, now with unrestricted access to all of Tycho's data. At first he tried the traditional Ptolemaic methods—epicycle, eccentric, and equant—but no matter how he varied the parameters, the calculated positions of the planet disagreed with Tycho's observations by up to 8 minutes of arc. His faith in the accuracy of Tycho's data led him to conclude that the Ptolemaic theory of epicycles, which had been used by Copernicus, would have to be replaced by a completely new theory, as he wrote: "Divine Providence granted us such a diligent observer in Tycho Brahe, that his observations convicted this Ptolemaic calculation of an error of eight minutes; it is only right that we should accept God's gift with a grateful mind.... Because those eight minutes could not be ignored, they alone have led to a total reformation of astronomy."

After eight years of intense effort Kepler was finally led to what is now known as the second of his two laws of planetary motion. The second law states that a radius vector drawn from the sun to a planet sweeps out equal areas in equal times, so that when the planet is close to the sun it moves rapidly and when far away it goes slowly. First correctly stated in book 5 of his *Epitome astrononiae Copernicanae* (*Epitome of Copernican Astronomy*), published in 1621, the law worked well for the earth's orbit, but when it was applied to Mars the 8-minute discrepancy once again appeared, which made Kepler realize that the planetary orbits might not be circular.

Kepler knew that the epicycle theory for Mercury gave an ovoid curve, but when he tried this for Mars the discrepancy was still 4 minutes of arc. Seeing that the ovoid curve he had drawn for the orbit of Mars was quite similar to an ellipse, he began thinking of the possibility that the planetary orbits were elliptical. As he wrote to his friend David Fabricius in July 1603: "I lack only a knowledge of the geometric generation of the oval or face-shaped curve.... If the figure were a perfect ellipse, then Archimedes and Apollonius would be enough." He soon realized that an ellipse would satisfy the calculations, but such an orbit did not fit in with his idea that the motion of the planets was driven by the magnetic field of the rotating sun. He expresses his frustration in chapter 58 of his *Astronomiae nova*, published in 1609:

> I was almost driven to madness in considering and calculating this matter. I could not find out why the planet would rather go on an elliptical orbit. Oh, ridiculous me! As the libration in the diameter could not also be the way to the ellipse. So this notion brought me up short, that the ellipse exists because of the libration. With reasoning derived from physical principles, agreeing with experience, there is no figure left for the orbit of the planet but a perfect ellipse.

Thus he arrived at what is now known as Kepler's first law of planetary motion, which is that each of the planets travels in an elliptical orbit, with the sun at one of the two focal points of the ellipse. The second law states that a radius vector drawn from the sun to a planet sweeps out equal areas in equal times. These two laws, which first appeared in Kepler's *Astronomia nova*, became the basis for his subsequent work on the *Rudolphine Tables*.

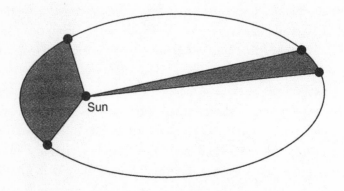

Kepler's first two laws of planetary motion: (1) each of the planets travels in an elliptical orbit, with the sun at one of the two focal points of the ellipse; (2) a radius vector drawn from the sun to a planet sweeps out equal areas in equal times

Kepler's first two laws of planetary motion eliminated the need for the epicycles, eccentrics, and deferents that had been used by astronomers from Ptolemy to Copernicus. The passing of this ancient cosmological doctrine was noted by Milton in book 8 of *Paradise Lost*:

> Hereafter, when they come to model Heaven,
> And calculate the stars; how they will wield
> The mighty frame; how build, unbuild, contrive
> To save appearances; how gird the sphere
> With centric and eccentric scribbled o'er,
> Cycle and epicycle, orb in orb.

Kepler wrote three other works on his research before the publication of his *Astronomia nova*. The first was *Astronomomiae pars optica*, published in 1603, and the second was *Ad Vitellionem paralipomena* (*Appendix to Witelo*), which came out the following year. Both books dealt with optical phenomena in astronomy, particularly parallax and refraction, as well as the annual variation in the size of the sun.

The third book was occasioned by another new star that appeared in October 1604 in the vicinity of Jupiter, Saturn, and Mars. Kepler published an eight-page tract on the new star in 1606 entitled *De stella nova*, with a subtitle describing it as "a book full of astronomical, physical, metaphysical, meteorological, astrological discussions, glorious and unusual." At the end of the tract Kepler speculated on the astrological significance of the new star, saying that it might be a portent of the conversion of the American Indians, a mass migration to the New World, the downfall of Islam, or even the second coming of Christ.

Meanwhile the whole science of astronomy had been profoundly changed by the invention of the telescope. Instruments called "perspective glasses" had been used in England before 1580 for viewing distant terrestrial objects, and both John Dee and Thomas Digges were known to be expert in their construction and use, though there is no evidence that they used them for astronomical observations. But their friend Thomas Harriot is known to have made astronomical observations in the winter of 1609–1610 with a small "telescope," which may have been a perspective glass.

Other than these perspective glasses, the earliest telescope seems to have appeared in 1604, when a Dutch optician named Zacharias Janssen constructed one from a specimen belonging to an unknown Italian, after which he sold some of them at fairs in northern Europe. When Galileo heard of the telescope he constructed one in his workshop in 1609, after which he offered it to the Doge of Venice for use in war and navigation. After improving on his original design, he began using his telescope to observe the heavens, and in March 1610 he published his discoveries in a little book called *Siderius nuncius* (*The Starry Messenger*).

Galileo sent a copy of the *Siderius nuncius* to Kepler, who received it on April 8, 1610. During the next eleven days Kepler composed his response in a little work called *Dissertatio cum Nuncio sidereal* (*Answer to the*

Sidereal Messenger), in which he expressed his enthusiastic approval of Galileo's discoveries and reminded readers of his own work on optical astronomy, as well as speculating on the possibility of inhabitants on the moon and arguing against an infinite universe. Galileo wrote a letter of appreciation to Kepler: "I thank you because you are the first one, and practically the only one, to have complete faith in my assertions."

Kepler borrowed a telescope from the elector Ernest of Cologne at the end of August 1610, and for the next ten days he used it to observe the heavens, particularly Jupiter and its moons. He published the results the following year in a booklet entitled *Narratio de Jovis satellitibus*, confirming the authenticity of Galileo's discovery.

That same year Kepler published a letter entitled *Strena*, which in its English translation is known as "A New Year's Gift, or On the Six-Cornered Snowflake," in which he reflects on the reason why snowflakes are hexagonal. Owen Gingerich described this minor work as "a perceptive, pioneering study of the regular arrangements and the close packing that are fundamental in crystallography."

Kepler's excitement over the possibilities of the telescope was such that he spent the late summer and early autumn of 1610 making an exhaustive study of the passage of light through lenses, which he published later that year under the title *Dioptrice*, which became one of the foundation stones of the new science of optics.

The death of Rudolph II in early 1612 forced Kepler to leave Prague and take up the post of district mathematician at Linz, where he remained for the next fourteen years. One of his official duties was a study of chronology, part of a program of calendar reform instituted by the archduke Ferdinand II, son of the late emperor Rudolph. As a result of his studies he established that Christ was born in what in the modern calendar would be 5 BC.

During the period that Kepler lived in Linz, he continued his calculations on the *Rudolphine Tables* and published two other major works, the first of which was *De harmonice mundi* (*The Harmony of the World*), which appeared in 1619. The title was inspired by a Greek manuscript of

Ptolemy's treatise on musical theory, the *Harmonica*, which Kepler acquired in 1607 and used in his analysis of music, geometry, astronomy, and astrology. The most important part of the *Harmonice* is the relationship now known as Kepler's third law of planetary motion, which he discovered on May 15, 1618, and presents in book 5. The law states that for each of the planets the square of the period of its orbital motion is proportional to the cube of its average distance from the sun.

There had been speculations about the relation between the periods of planetary orbits and their radii since the time of Pythagoras, and Kepler was terribly excited that, following in the footsteps of Ptolemy, he had at last found the mathematical law "necessary for the contemplation of celestial harmonies." He wrote of his pleasure "that the same thought about the harmonic formulation had turned up in the minds of two men (though lying so far apart in time) who had devoted themselves entirely to contemplating nature.... I feel carried away and possessed by an unutterable rapture over the divine spectacle of the heavenly harmony."

Kepler dedicated the *Harmonice* to James I of England. The king responded by sending his ambassador, Sir Henry Wooton, with an invitation for Kepler to take up residence in England. But after considering the offer for a while, Kepler decided against it.

The English poet John Donne was familiar with the work of Copernicus and Kepler, probably through Thomas Harriot. Donne had in 1611 said to the Copernicans that "those opinions of yours may very well be true ... creeping into every man's mind." That same year Donne lamented the passing of the old cosmology in "An Anatomy of the World":

> And new Philosophy calls all in doubt,
> The Element of fire is quite put out;
> The Sun is lost, and th'earth, and no man's wit
> Can well direct him, where to look for it.

Kepler's second major work at Linz was his *Epitome astronomiae Copernicanae* (*Epitome of Copernican Astronomy*), published in 1621. In the first three of the seven books of the *Epitome* Kepler refutes the traditional arguments against the motions of the earth, going much farther than Copernicus and using principles that Galileo would later give in greater detail. His three laws of planetary motion are explained in great detail in book 4, along with his lunar theory. The last three books treat practical problems involving his first two laws of planetary motion as well as his theories of lunar and solar motion and the precession of the equinoxes. The work became very popular, and, according to J. L. Russell, "from 1630 to 1650 the *Epitome* was the most widely read treatise on theoretical astronomy in Europe."

In 1626 Kepler was forced to leave Linz and move to Ulm, from where he moved on to Sagan in 1628, then two years later he traveled to Regensburg, where he died of an acute fever on November 15, 1630. His tombstone, now lost, was engraved with an epitaph that he had written himself: "I used to measure the heavens, now I measure the shadow of the earth. / Although my soul was from heaven, the shadow of my body lies here."

While Kepler was still in Ulm he published the *Rudolphine Tables* in September 1627, dedicating them to the archduke Ferdinand II. A decade earlier Kepler had come across the pioneering work on logarithms published in 1614 by the Scottish mathematician John Napier (1550–1617). He used logarithms in computing the planetary positions in the *Rudolphine Tables*, appending logarithmic tables as well as a catalogue of the celestial coordinates of a thousand stars, all based on the Copernican theory. The new tables were far more accurate than any in the past, and they remained in use for more than a century.

Kepler used his tables to predict that Mercury and Venus would make transits across the disk of the sun in 1631. The transit of Venus was not observed in Europe because it took place at night. The transit of Mercury was observed by Pierre Gassendi in Paris on November 7, 1631, representing a triumph for Kepler's astronomy, for his prediction was in error by only 10 minutes of arc as compared to 5° for tables based on Ptolemy's model.

Kepler's last work, published posthumously in 1634, was a curious tract called *Somnium seu astronomia lunari* (*Dream on Lunar Astronomy*), which has been described as the first work of science fiction. It tells of how a student of Tycho Brahe is transported by occult forces to the moon, from where he describes the earth and its motion around the sun along with the other planets.

Kepler's three laws of planetary motion were to become the basis of the new heliocentric astronomy that emerged in the seventeenth century, culminating in Newton's *Principia*, published in 1687. When the *Principia* was first introduced to the Royal Society in London, it was described as "a mathematical demonstration of the Copernican hypothesis as proposed by Kepler." The astronomer Edmond Halley, in his review, wrote that Newton's first eleven propositions in book 1 of the *Principia* were "found to agree with the Phenomena of the Celestial Motions, as discovered by the great Sagacity and Diligence of Kepler."

13

The Great Debate

*W*HILE KEPLER ROSE TO FAME EXAMINING PLANETARY MOTIONS, Galileo had been quietly pursuing his researches in the science of motion, until he was forced to interrupt them to concentrate his efforts on advancing the Copernican theory, which soon embroiled him in controversy.

Galileo was born in Pisa on February 15, 1564, to a patrician Florentine family that moved back to Florence in 1575. His father was Vincenzio Galilei, a musician and musical theorist, who in 1589 published a work on the numerical theory of musical harmony, based on an experimental study of consonance and its relation to the lengths and tensions of strings. Galileo's interest in the testing of laws of mathematical physics by observation probably stems from his father's musical investigations.

Already a precocious young student, Galileo received his elementary education from a private tutor before being sent off to school at the renowned monastery of Santa Maria in Vallombrosa. He was enrolled in the school of medicine at the University of Pisa in 1581, studying physics and astronomy under Francesco Buonamici, philosophy with Girolamo Borro, and mathematics with Filippo Fanton. He left Pisa without a degree in 1585 and returned to Florence, where he began an independent study of Euclid and Archimedes under Ostilio Ricci.

During the years 1585–1589 Galileo gave private lessons in mathematics in Florence as well as private and public instruction in Siena.

In 1583 Galileo had already made his first scientific discovery, that the period of a pendulum is independent of the angle through which it swings, at least for small angles. Supposedly, he came to this conclusion by observing the oscillation of a chandelier in a cathedral. Three years later he invented a hydraulic balance, which he described in his first scientific publication, *La balancetta* (*The Little Balance*), based on Archimedes' principle, which he also used in determining the center of gravity of solid bodies.

Galileo Galilei

Galileo was appointed professor of mathematics in 1589 at the University of Pisa, where he remained for only three years. During this period he wrote an untitled treatise on motion now referred to as *De motu* (*On Motion*), which remained unpublished during his lifetime. The treatise was an attack on Aristotelian physics, such as the notion that heavy bodies fall more rapidly than light ones, which Galileo is supposed to have refuted by dropping weights from the Leaning Tower of Pisa, though there is reason to believe that this was merely a "thought experiment."

A similar experiment had already been performed in 1586 by the Flemish engineer Simon Stevin (1548–1620), court mathematician to Prince Maurice of Orange. Stevin took two lead balls, one ten times the weight of the other, and dropped them thirty feet onto a wooden board from the church tower in Delft, concluding from the sound that they reached the ground at the same time, thus refuting Aristotle's theory.

Stevin was the first to establish the basic laws of hydrostatics and statics, in both cases beginning with the assumption that perpetual motion is impossible. He showed that the pressure of a liquid on the base of the containing vessel depended only on depth and was independent of shape and volume, the basic law of hydrostatics. He proved the basic law of statics, the equilibrium of forces, by a demonstration clearly described by Crombie:

> With the same assumption of the impossibility of perpetual motion
> he showed also why a loop of cord, on which weights were attached
> at equal distances apart. He showed that as long as the bottom of
> the prism was horizontal no movement occurred in the upper sec-
> tion of the cord when the suspended section was removed, and
> from this he arrived at the conclusion that weights on inclined
> planes were in equilibrium when proportional to the lengths of
> their supporting planes cut by the horizontal…. This conclusion
> implied the idea of the triangle or parallelogram of forces, which
> Stevin applied to more complicated machines.

The principal mathematical work of Stevin is his *De thiende* (*The Art of Tenths*), first published in Dutch in 1585 and translated into English as *Decimal Arithmetic: Teaching how to perform all computations whatsoever by whole numbers without fractions, by the four principles of common arithmetic: namely, addition, subtraction, multiplication, and division*. Decimal fractions had been in use since antiquity, but they did not become widely known until the publication of Stevin's book. He thought that the usefulness of decimal fractions was such that it was only a matter of time before the decimal system became the universal basis for coinage as well as weights and measures.

His other writings include a pioneering work on musical theory, published in 1585. This was apparently inspired by the writings of Vincenzo Galilei, Galileo's father. Galileo in turn would be inspired by Stevin's writings, a remarkable cycle that says much about the developing character of science during the Renaissance.

Stevin also wrote on other areas of science, such as optics, astronomy, and geography, some of which were translated into Latin by the Dutch astronomer and mathematician Willebrord Snellius (1580–1626). Snellius was professor of mathematics at the University of Leiden. In 1617 he published a work called *Eratosthenes Batavus* (*The Dutch Eratosthenes*), in which he reported his determination of the circumference of the earth by measuring the distance between two towns separated by 1° of latitude, with an error of about 3 percent compared to the modern value. Snellius was the first to formulate the law of refraction mathematically, using the sine function. He made this discovery in 1621, but he never published it. It is possible that he was led to this discovery when he translated Stevin's work on optics.

The most distinguished student that Snellius taught at Leiden was the Dutch scientist and philosopher Isaac Beeckman (1588–1637). Beeckman became rector of the Latin school in Dordrecht, where he founded a *Collegium Mechanicum*, or Technical, investigating natural phenomena, which he recorded in his *Journal*. Prior to 1620 he befriended Simon Stevin, who at the time was court mathematician to Prince Maurice of Orange. He also became a close friend of René Descartes (1596–1650), who in those years was a young soldier in the army of Prince Maurice and already a highly regarded mathematician. Descartes was deeply influenced by Beeckman's ideas on motion, though he later denied this. Others who were influenced by Beeckman were Marin Mersenne (1588–1648) and Gassendi (1592–1655), both of whom were important figures in the seventeenth-century Scientific Revolution.

Galileo's *De motu* was probably written in opposition to a massive work of the same title by his teacher Francesco Buonamici, who had first introduced him to the study of motion, as scholar Mario Helbing pointed out:

> Buonamici probably introduced Galileo to the atomism of Democritus, to Philoponus's critiques of Aristotle's teachings, to Copernicus's innovations in astronomy, to Archimedes and his use of the buoyancy principle to explain upward motion, to Hipparchus's theory of impetus, and to the writings of Clavius and Benedetto Pereira at the Collegio Romano—references to all of which can be found in his *De Motu*.

Although Galileo rejected many of Buonamici's ideas, which were Aristotelian, he was deeply influenced by his teacher and was in substantial agreement with him in two areas, as Helberg noted:

> The first is the general methodology they employ in their theory of motion. Both wish to employ a *methodus* to put their science on an axiomatic base, imitating in this the reasoning process of mathematicians. Both regard sense experience as the foundation of natural science, taking this in a sense broad enough to include experiment in the rudimentary form it was then assuming at Pisa. And both see causal reasoning and demonstration, with its two-fold process of resolution and composition, as the normal road to scientific conclusions. The second and most important area of agreement is the status each accords to mathematics, both as a science in its right and as an aid in investigating the secrets of nature.

While teaching at Pisa, Galileo was influenced by the writings of Giovanni Battista Benedetti (1530–1590), who was highly critical of Aristotle's ideas on motion. One of the Aristotelian concepts he rejected was that heavier bodies fall faster than lighter ones, which he refuted by a "thought experiment," summarized by Crombie, who pointed out the error that Galileo made in drawing his conclusion from this demonstration:

He imagined a group of bodies of the same material and weight falling beside each other, first connected and then separately, and he concluded that their being in connection could not alter their velocity. A body the size of the whole group would, therefore, fall with the same velocity as each of its components. He therefore concluded that all bodies of the same material (or "nature"), whatever their size, would fall with the same velocity, though he made the mistake of believing that the velocities of bodies of the same volume but different material would be proportional to their weights.

Crombie went on to discuss three other ideas proposed by Benedetti, two of which originated with the Greeks and the third and most interesting with Leonardo da Vinci:

> Inspired by Archimedes, he thought of weight as proportional to the relative density in a given medium. He then used the same argument as Philoponus to prove that velocity would not be infinite in a void. Benedetti also held that in a projectile natural gravity was not entirely eliminated by the *impetus* of flight, and he followed Leonardo in maintaining that impetus engendered movement only in a straight line, from which it might be deflected by a force, such as the "centripetal" force exerted by a string, which prevented a string swung in a circle flying off at a tangent.

Galileo, in his *De motu*, rejected the Aristotelian theory in which bodies fall or rise to their natural place, with earth at the center and above it in succession water, air, and fire. Instead, he said that whether a body moves downward or upward depends on its density relative to that of the medium in which it is moving. He then argued that these phenomena should be seen as balance problems, so that the ninth section in *De motu* is entitled "In which all that was demonstrated above is considered in physical terms, and bodies moving naturally are reduced to weights on a balance," a method he credited to Archimedes. Referring to two weights e and o suspended from opposite ends of an equal-arm balance, he wrote:

> Now in the case of weight e there are three possibilities: it may either be at rest, or move upward, or move downward. If therefore weight e is heavier than weight o, then e will move downward. But if e is less heavy it will, of course, move upward, and not because it does not have weight, but because the weight of o is greater. From this it is clear that, in the case of the balance, motion upward as well as downward takes place because of weight but in a different way. For motion upward will occur for e on account of the weight of o, but motion downward will occur for e on account of its own weight. But if the weight of e is equal to that of o, then e will move neither upward nor downward.

Thus Galileo proposed that a method for solving all problems of motion, including those of floating bodies, could be reduced to that of weights on an Archimedian balance. He went on to show that all simple machines, such as the lever, the inclined plane, and the pendulum, could also be reduced to balance problems. The problem of free fall he interpreted as an instance of floating bodies or a balance that had no weight on one side.

In 1592 Galileo was appointed to the chair of mathematics at the University of Padua, where he remained for eighteen years. During his time at Padua he began living with a Venetian woman named Marina, by whom he fathered two daughters, Virginia and Livia, both of whom were put in a nunnery at an early age, and a son, Vincenzio, whom he later legitimized.

During his Paduan residency Galileo wrote several treatises for the use of his students, including one that was first published in a French translation in 1634 under the title *Le meccaniche*, a study of motion and equilibrium on inclined planes that further developed the ideas he had presented in *De motu*. Here he bridges the gap between statics and dynamics by remarking that an infinitesimal force would be sufficient to disturb equilibrium.

In *Le mecchanice* Galileo used the concept of center of gravity to discuss simple machines, as when he said, "Thus, *moment* [momento] is the impetus to go downward composed of heaviness, position and of anything else by which this tendency may be caused." The model that he employed is a single-arm balance or lever, where moment is a generalized force. In the case of a balance, moment is the product of weight times the perpendicular distance between the line of action of the weight and the fulcrum, while for a body moving on an inclined plane the moment depends on the angle of inclination of the plane. Using this model, he went on to explain the lever and other types of machines before finally making a first attempt to deal with the force of impact, all in the Archimedean tradition. Galileo would use this model throughout his career, including in his last two great works, the *Dialogue Concerning the Chief World Systems, Ptolemaic and Copernican*, and the *Discourses and Mechanical Demonstrations Concerning Two New Sciences, of Mechanics and of Motions*.

In May 1597 Galileo wrote to a former colleague at Pisa defending the Copernican theory. Three months later he received a copy of *Mysterium cosmographicum*, which led to his first correspondence with Kepler.

Around the same time Galileo constructed a mathematical instrument that he called the geometric and military compass, later called the sector, which served among other purposes that of the instrument now known as the proportional divider. He hired a skilled artisan to make and sell this and other instruments in his own workshop. In 1606 he published his first work, a handbook of instructions for users of the compass, dedicating it to the young prince Cosimo de' Medici, to whom he had been giving private lessons.

In 1604 Galileo wrote a letter to his friend Paolo Sarpi, saying that on the basis of an axiom he had proved that the distances covered by a falling body are proportional to the square of the times. This is the law of kinematics now written as $s = 1/2at^2$, where s is the distance, a is the acceleration, and t is the time. The axiom that he adopted was that, for a body falling from rest, the instantaneous velocity, that is, the velocity at any moment in time, is proportional to the distance traversed, which is not true.

The fact that the acceleration was proportional to the time rather than the distance had first been stated unequivocally by the Spanish Dominican friar Domingo de Soto (1494–1560). Thus Galileo had derived the correct law giving the distance covered as a function of time for a body moving under the influence of gravity, though he based his derivation on an erroneous assumption. Galileo later corrected this mistake, apparently after carrying out an experiment, as Crombie pointed out:

> It seems likely that Galileo had discovered his mistake and correctly formulated the law of acceleration and the space theorem by 1609, although he published them only in the *Two Principal Systems* in 1632. It is possible that he had already carried out his experiment to test the law with a bronze ball rolling down an inclined plane as early as 1604. This experiment is described in the Two New Sciences (1638), where the mathematical demonstration is again set out. In the absence of an accurate clock, he defined equal intervals of time as those during which equal weights of water issued from a small hole in a bucket; he used a very large amount of water relative to the amount issuing through the hole, so that the decrease in head was unimportant. His experiment confirmed his definition and law of free fall, and from it he deduced further theorems.

The year 1609 was a turning point in Galileo's career, for it was then that he first learned of the existence of the telescope. He quickly made one for himself, eventually developing an improved model with a magnifying power of thirty. The following year he described his observations in the *Siderius nuncius* (*The Starry Messenger*), dedicated to Cosimo de' Medici, who by then had succeeded as Grand Duke of Tuscany.

After the dedication Galileo gives a brief summary of the book, which in its English translation reads: "ASTRONOMICAL MESSAGE/Which contains and explains recent observations made with the aid of a new spyglass concerning the surface of the moon, the Milky Way, nebulous stars, and innumerable fixed stars, and four planets never before seen, and now named THE MEDICEAN STARS."

The "Medicean Stars" were the four principal moons of Jupiter that Galileo discovered with his telescope. He called them "planets" because they circled Jupiter just as it and the other planets orbit the sun, as he writes in the dedication, the first public notice by Galileo that he accepted the Copernican system:

> Behold, then, four stars to bear your famous name; bodies which belong not to the incongruous multitude of fixed stars, but to the bright ranks of the planets. Variously moving about most noble Jupiter as children of his own, they complete their orbits with marvelous velocity—at the same time executing with one harmonious accord mighty revolutions every dozen years about the center of the universe, that is, the sun.

Galileo begins the book with an account of his systematic observation of the moon, which he describes in great detail and illustrates with drawings and geometrical diagrams. As he writes of his first lunar observation: "On the fourth or fifth day after new moon, when the moon is seen with brilliant horns, the boundary which divides the dark part from the light does not extend uniformly in an oval line as would happen on a perfectly spherical solid, but traces out an uneven, rough and very wavy line."

He concluded that "the surface of the moon is not smooth, and precisely spherical as a great number of philosophers believe it (and the other heavenly bodies) to be, but is uneven, rough, and full of cavities and prominences, being not unlike the face of the earth, relieved by chains of mountains and deep valleys."

He then writes of his survey of the planets and stars, noting that with his telescope the planets were magnified and appeared as illuminated globes, looking like small moons, while the fixed stars still appeared as brilliant points of scintillating light, though much brighter than when viewed with the naked eye:

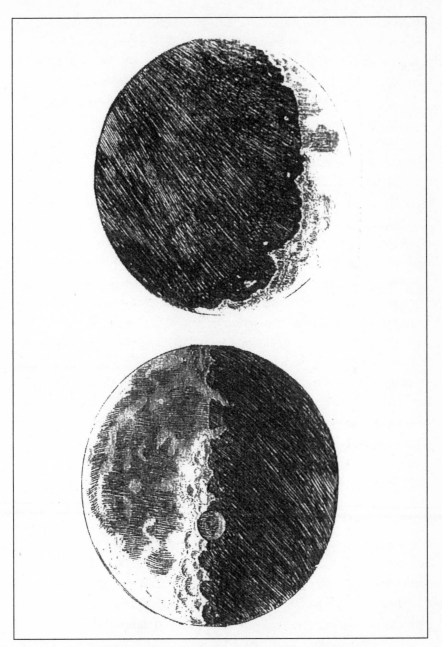

The surface of the Moon, from Galileo's Sidereus nuncius

Deserving of notice also is the difference between the appearance of the planets and of the fixed stars. The planets show their globes perfectly round and definitely bounded, looking like little moons, spherical and flooded all over with light; the fixed stars are never seen to be bounded by a circular periphery, but have rather the aspect of blazes whose rays vibrate about them and scintillate a great deal. Viewed with a telescope they appear of a shape similar to that which they present to the naked eye, but sufficiently enlarged so that a star of the fifth or sixth magnitude seems to equal the Dog Star, the largest [actually not the largest but the brightest] of all the fixed stars.

Galileo saw that in addition to stars of the sixth magnitude, the faintest of those that can be seen with the naked eye, "a host of other stars are perceived through the telescope which escape the naked eye; these are so numerous as almost to surpass belief. One may, in fact, view all the stars included among the first six magnitude."

He observed the constellation Orion, intending to depict the new stars he had observed along with the familiar ones visible to the naked eye, but he "was overwhelmed by the vast quantity of stars and by the limitations of time, so I have deferred this to another occasion. There are more than five hundred new stars distributed among the old ones within limits of one or two degrees of arc." Thus, as he writes, he limited the scope in his illustration:

Hence to the three stars in the Belt of Orion and the six in the Sword which were previously known, I have added eighty adjacent stars discovered recently, preserving the intervals between them as exactly as I could. To distinguish the known or ancient stars, I have depicted them larger and outlined them doubly; the other (invisible) stars I have drawn smaller and without the extra line. I have also preserved differences of magnitude as well as possible.

He did the same with the Pleiades, where in addition to the six brightest of the seven visible stars of the constellation he depicted thirty-six new ones in the illustration. He then turned his attention to the Milky Way, where, as he notes, he resolved questions about its nature and composition that had been argued since antiquity:

> With the aid of the telescope this has been scrutinized so directly and with such ocular certainty that all the disputes which have vexed philosophers through so many ages have been resolved, and we are at last freed from wordy disputes about it. The galaxy is, in fact, nothing but a congeries of innumerable stars grouped together in clusters. Upon whatever part of it the telescope is turned a vast crowd of stars is immediately presented to view. Many of them are rather large and quite bright, while the number of smaller ones is quite beyond calculation.

Besides the Milky Way, he observed other celestial "clouds" that in the past had been "nebulous," and he found that they too turned out to be clusters of previously invisible stars. He depicted two of these star clusters in his illustrations, the Nebula of Orion and the Nebula of Praesepe, which lies in the constellation Cancer.

Galileo made an improved version of his telescope with a higher magnification, and with this, on January 7, 1610, he discovered what at first he thought to be small stars close to Jupiter and aligned with it, two on the eastern side and one on the west. The following night he found that the three "starlets," as he called them, were all to the west of Jupiter and aligned at equal intervals from one another and the plane. Observations on subsequent nights revealed the starlets changing their orientations with respect to one another and to Jupiter, though still forming a straight line. On January 10 only two of them appeared, both of them to the east of Jupiter, leading Galileo to suppose that the third was hidden behind the planet. On January 13 he saw for the first time four starlets, one westerly and the others easterly, the middle western star slightly to the north of the others. He eventually concluded that the four starlets were in fact satellites of Jupiter, those closest to the planet revolving around it more swiftly than those farther away, just as the planets orbit the sun, which he presented as a conclusive argument in favor of the Copernican system:

> Here we have a fine and elegant argument for quieting the doubts of those who, while accepting with quiet tranquility the revolutions of the planets about the sun in the Copernican system, are mightily disturbed to have the moon alone revolve around the earth and accompany it in an annular rotation around the sun. Some have believed that this structure of the universe should be rejected as impossible. But now we have not just one planet rotating about another while both run through a great orbit around the sun; our own eyes show us four stars which wander around Jupiter as does the moon around the earth, while all together trace out a grand revolution about the sun in the space of twelve years.

News of Galileo's discoveries spread throughout Europe and the wider world, bringing him great fame even beyond the scientific community. Just five years after the *Siderius nuncius* first appeared, a summary of Galileo's discoveries was published in Chinese by a Jesuit missionary in Peking, the first penetration of the new astronomy into the Orient.

Cosimo de' Medici responded by making Galileo his court philosopher and appointing him to the chair of mathematics at the University of Pisa. Galileo had no obligation to teach at the University of Pisa or even to reside in the city, and so after his appointment, in September 1610, he departed to take up residence in Florence.

Meanwhile Galileo had made two more important discoveries with his telescope. The first was the curious shape of Saturn, which appeared to have a pair of equatorial protuberances. These were actually the famous rings of Saturn, which his primitive telescope was unable to resolve, and he thought that they were a pair of small satellites in stationary orbits close to the planet. The second discovery was that Venus goes through a cycle of phases similar to those of the moon, the only difference being that its size varied as its distance from the earth changed. Galileo showed that this was due to the fact that Venus was in orbit around the sun rather than the earth. He communicated these discoveries to Kepler and to the Jesuit astronomers in Rome as added evidence in support of the Copernican theory.

Late in 1610 Galileo received a congratulatory letter from the eminent Jesuit astronomer Father Christopher Clavius, chief mathematician at the Roman College, informing him that the astronomers there had verified his discovery of the new fixed stars and the satellites of Jupiter. Thus encouraged, in March of the following year Galileo visited Rome on what a contemporary described as a triumphal tour, in which he was welcomed by noblemen and by church dignitaries and was given a friendly interview by Pope Paul V.

During Galileo's stay in Rome he was elected to the Accademia dei Lincei (Academy of the Lynx-Eyed), founded in 1603 by Federigo Cesi, Marquis of Montecelli. This was the first of the European societies specializing in philosophy and science, which would play such an important role in the Scientific Revolution, the most renowned being the Accademia Cimento (1657–1567), the Royal Society of London, which received its first charter in 1662, and the Académie Royale des Sciences, which held its first meetings in 1666.

Cardinal Robert Bellarmine, head of the Roman College, asked the mathematicians on the faculty to evaluate Galileo's discoveries. When they confirmed them, Clavius and his colleagues honored Galileo with a full day of ceremonies at the college, during which one of them delivered an oration praising him and his book.

But, as Galileo biographer Stillman Drake wrote, "About the time Galileo left Rome to return to Florence, a letter went secretly to the chief inquisitor at Padua upon instructions from Bellarmine and six of his cardinals. It contained the ominous words: 'see if Galileo is mentioned in the proceedings against Dr. Cesare Cremonino.'" Cremonino, head of the Philosophy Department at the University of Padua, was a good friend of Galileo, despite their philosophical differences. He was long suspected of heretical opinions by the inquisitors, but he was never brought to trial because he remained in Venetian territory.

Despite the confirmation of the Jesuit astronomers in Rome, conservative Aristotelians at the universities remained skeptical of Galileo's discoveries. Giulio Libri, who taught philosophy both at Pisa and Padua while Galileo was at those universities, refused even to look through a telescope. When Libri died in 1610, Galileo expressed the hope that since he had refused to look at the celestial bodies through a telescope when he was on earth, he could now see them with his own eyes on the way to heaven.

Galileo had resumed his earlier researches on hydrostatics in 1609, and three years later he published, in Italian, a treatise entitled *Discourse on Floating Bodies*. Here, as Stillman Drake noted, "Using the concept of moment and the principle of virtual velocities, Galileo extended the scope of the Archimedean work beyond purely hydrostatic considerations." This was a continuation and extension of his earlier researches in hydrostatics, to which he had returned partly because of a dispute concerning floating bodies in which he was involved with Lodovico delle Colombe, leader of a group of Aristotelians who were highly critical of his efforts in support of the Copernican theory.

Meanwhile Galileo had become involved in another dispute over a work by a German astronomer named Father Christopher Scheiner, a Jesuit professor at the University of Ingolstadt, who had observed spots on the sun. Scheiner's superior had forbidden him to publish his findings under his own name, and so he presented them in the form of several letters addressed to his friend Mark Welser, a wealthy merchant of Augsburg. Welser sent the letters to Galileo, with whom he had corresponded earlier concerning the lunar mountains, asking his opinion, concealing Scheiner's name under the pseudonym "Apelles." Galileo responded with a series of three letters, written in Italian since that was the language in which Welser had addressed him. Scheiner had the first of Galileo's letters translated into German, after which he wrote a reply entitled *A More Accurate Discussion of Sunspots and the Stars Which Move Around Jupiter*, to which Galileo replied in his third letter to Welser. Scheiner's first recorded observation of sunspots was made on October 21, 1611. Galileo's earliest known mention of the phenomenon occurs in a letter dated October 1621, in which he indicates that he had observed sunspots eighteen months earlier and had shown them to others while in Rome. He was apparently unaware that the Englishman Thomas Harriot had already reported his observation of sunspots, as had the German Johannes Fabricius, who announced his discovery in a booklet printed in the summer of 1611.

In all three of his letters Galileo gave detailed evidence countering Scheiner's theory that the spots were actually tiny planets orbiting close to the sun. Then, after giving an extremely detailed description of the phenomenon, he presented his own, correct, view that sunspots were part of the fluid solar surface and were in rotation about the sun's axis along with the rest of the solar sphere.

In the first letter Galileo also gave a detailed description of the moons of Jupiter as well his two more recent discoveries, that is, the phases of Venus and the protuberances on Saturn's disk, which he thought to be small moons. Then in his third letter he presented these discoveries as additional evidence for the Copernican theorem, predicting that it would soon be universally adopted. As he wrote, referring to the protuberances on Saturn: "I say, then, that I believe that after the winter solstice of 1614 they may once more be observed. And perhaps this planet also, no less than horned Venus, harmonizes admirably with the great Copernican system, to the universal revelation of which doctrine propitious breezes are now seen to be directed toward us, leaving little fear of clouds or crosswinds."

The second letter also contains Galileo's first published mention of the concept of inertia, according to which a body will preserve a state of uniform linear motion or of rest unless acted upon by a force. He states this principle in discussing the rotation of sunspots on the solar surface: "For I seem to have observed that physical bodies have physical inclination to some motion (as heavy bodies downward), which motion is exercised by them through an intrinsic property and without need of a particular external mover, whenever they are not impeded by some obstacle."

Galileo's *Letters on Sunspots*, containing his entire correspondence on the subject with Welser, was printed at Rome in 1613 under the auspices of the Lincean Academy. Since it was in Italian, as Stillman Drake remarked, it "thus brought the question of the earth's motion to the attention of virtually everyone in Italy who could read."

Meanwhile Galileo had been active in advancing the cause of Copernicanism against the accepted cosmology of Aristotle, which in its reinterpretation by Saint Thomas Aquinas formed part of the philosophical basis for Roman Catholic theology. This led to severe criticism of Galileo by his enemies, who interpreted his arguments as being attacks against the Church itself. On December 16, 1611 Galileo's friend Lodovico Cigoli wrote to him from Rome about a conspiracy to denounce him publicly:

I have been told by a friend of mine, a priest who is very fond of you, that a certain crowd of ill-disposed men envious of your virtue and merits meet at the house of the archbishop there and put their heads together in a mad quest for any means by which they could damage you, either with regard to the motion of the earth or otherwise. One of them wished to have a preacher state from the pulpit that you are asserting outlandish things. The priest having perceived the animosity against you, replied as a good Christian and a religious man ought to do. Now I write this to you so that your eyes will be open to such envy and malice on the part of evildoers.

The climax of these attacks came on December 21, 1614, when the Dominican priest Thomas Caccini, speaking from the pulpit of his church in Florence on the book of Joshua, denounced Galileo, the Copernican system, and mathematics in general as being contrary to Christianity. Galileo's response came in the form of a letter to the Grand Duchess Christina, mother of Cosimo II de' Medici, who had expressed interest in his views. This work was probably completed by June 1615, though it was not published until 1636, under the title *Letter to Madame Christina of Lorraine, Grand Duchess of Tuscany, Concerning the Use of Biblical Quotations in Matters of Science*. Here he argued that sacred scripture had to be interpreted in terms of what science knew about the world. Neither the Bible nor nature could speak falsely, he said, but the investigation of nature was the duty of scientists, and theologians should reconcile scientific facts with the language of the scriptures, noting that the Bible "was not written to teach us astronomy." His gives the book of Joshua as an example of how a scriptural statement could be better interpreted in terms of the Copernican theory than with the Aristotelian geocentric model.

> But, if I am not mistaken, something of which we are to take no small account of is that by the aid of this Copernican system we have the literal, open, and easy sense of another statement that we read in the same miracle, that the sun stood still *in the midst of the heavens* ... they [the theologians] are forced to interpret the words *in the midst of the heavens* a little knottedly, saying that this is no more than that the sun stood still while it was in our hemisphere, that is, above our horizon. But unless I am mistaken, we can avoid this and all other knots if, in agreement with the Copernican system, we place the sun in the "midst"—that is, in the center—of the celestial orbs and planetary rotation, as it is most necessary to do. Then take any hour of the day, either noon, or any hour as close to evening as you please, and the day would be lengthened and all the celestial revolutions stopped by the sun's standing still *in the midst of the heavens*; that is, in the center, where it resides. This sense is much better accommodated to the words.... For the true and only "midst" of a spherical body such as the sky is its center.

Pope Paul V, who was disturbed by the discussion of biblical interpretation, at the time a serious issue with the Protestants, appointed a commission to study the question of the earth's motion. The commission decided that the teachings of Copernicus were probably contrary to the Bible, and on February 28, 1616, Galileo was told by Cardinal Bellarmine that he could no longer hold or defend the heliocentric theory. On March 3 Bellarmine reported that Galileo had acquiesced to the pope's warning, and that ended the matter for the time being.

On March 5, 1616, the Holy Office of the Inquisition in Rome placed the works of Copernicus and all other writings that supported it on the Index, the list of books that Catholics were forbidden to read, including those of Kepler. The decree held that believing the sun to be the immovable center of the world is foolish and absurd, philosophically false, and formally heretical. Pope Paul V instructed Cardinal Bellarmine to censure Galileo, admonishing him not to hold or defend Copernican doctrines any longer.

After his censure Galileo returned to his villa in Arcetri outside Florence, where for the next seven years he remained silent. But then in 1623, after the death of Paul V, Galileo took hope when he learned that his friend Maffeo Cardinal Barberini had succeeded as Pope Urban VIII. Heartened by his friend's election, Galileo proceeded to publish a treatise entitled *Il saggiatore* (*The Assayer*), which appeared later that year, dedicated to Urban VIII.

Il saggiatore grew out of a dispute over the nature of comets between Galileo and Father Horatio Grassi, a Jesuit astronomer. This had been stimulated by the appearance in 1618 of a succession of three comets, the third and brightest of which remained visible until January 1619. Grassi, writing anonymously, took the view of Tycho Brahe that the comets were phenomena occurring in the celestial regions, using this as an argument against the Copernican theory in favor of the Tychonic model, in which Mercury and Venus orbit the sun, which, with the other visible planets, orbit the immobile earth. Galileo, who was bedridden at the time, discussed the comets with his disciple Mario Giudecci, who gave two lectures on the subject at the Florentine Academy. The lectures were published in Giudecci's name, though in his introductory remarks he acknowledged that the ideas he presented were largely those of Galileo. In these lectures the ideas of the anonymous Jesuit were criticized, particularly concerning his support of the Tychonic model rather than the Copernican theory. Grassi, writing under the pseudonym of Lotario Sarsi, replied with a direct attack on Galileo's views in a book, published later in 1619. After a long delay, Galileo responded with *Il saggiatore*, in the form of a letter to his disciple Virginio Cesarini, published by the Accademia dei Lincei in 1623.

Since Galileo could no longer defend the Copernican theory, in *Il saggiatore* he avoided the contentious question of the earth's motion, but rather presented a general scientific approach to the study of celestial phenomena. Responding to the charge that he had criticized the ideas of an established authority like Tycho Brahe, he replied, "In Sarsi I seem to discern the firm belief that in philosophizing one must support oneself upon the opinion of some celebrated author, as if our minds ought to remain completely sterile and barren unless wedded to the reasoning of some other person." He went on to say that philosophy is not a book of fiction such as the *Iliad* or *Orlando Furioso*, where "the least important thing is whether what is written is true," after which he made his famous statement that the book of nature is written in mathematical characters.

> Well, Sarsi, that is not how matters stand. Philosophy is written in this grand book, the universe, which stands continually open to our gaze. But the book cannot be understood unless one learns to comprehend the language and read the letters in which it is composed. It is written in the language of mathematics, and its characters are triangles, circles and other geometrical figures, without which it is humanly impossible to understand a single word of it, without these, one wanders about it in a dark labyrinth.

Il saggiatore was favorably received in the Vatican, and Galileo went to Rome in the spring of 1623 and had six audiences with the pope. Urban praised the book, but he refused to rescind the 1616 edict against the Copernican theory, though he said that if it had been up to him the ban would not have been imposed. Galileo did receive Urban's permission to discuss Copernicanism in a book, but only if the Aristotelian-Ptolemaic model was given equal and impartial attention.

Encouraged by his conversations with Urban, Galileo spent the next six years writing a book called the *Dialogue Concerning the Chief World Systems, Ptolemaic and Copernican*, which was completed in 1630 and finally published in February 1632. The book is divided into four days of conversations between three friends: Salviati, a Copernican, who is a spokesman for Galileo himself; Simplicio, an Aristotelian, named for the famous commentator on Aristotle, Simplicius; and Sagredo, an educated layman, named for Galileo's late friend, Giovan Francesco Sagredo, whom each of the other two tries to convert to his point of view.

The first day is devoted to a critical examination of Aristotelian cosmology, particularly those having to do with the distinction between the terrestrial and celestial regions, and the notion that the earth is the immobile center of the universe, both of which are rejected. However, the Aristotelian theory that the natural motion of celestial bodies is circular was accepted, but Galileo argues that circular motion is also natural for bodies on the rotating earth. He also rejected the Aristotelian notion of motion as a process requiring a continuous cause, and instead he gave a definition that would enable him to measure it, saying, "Let us call velocities equal, when the spaces passed have the same proportion as the times in which they are passed."

On the second day the objections against the earth's motions are refuted on physical grounds. Many of Galileo's arguments here, though persuasive, are based on his erroneous notion that circular motion is natural for bodies on the rotating earth. Galileo does say in his preface that any experiment performed would give the same result whether the earth is in motion or at rest. Concerning the Aristotelian theory of gravitation, Galileo has Sagredo give this response to Simplicio's statement that everyone knows that what causes bodies to fall downward is gravity:

> You are wrong, Simplicio; you should say that everyone knows that it is called gravity. But I am not asking you for the name but the essence of the thing. Of this you know not a bit more than you know the essence of the stars in gyration. I except the name that has been attached to the former and domestic by the many expressions we have of it a thousand times a day. We don't really understand what principle or what power it is that moves a stone downwards, any more than we understand what moves it upward after it has left the projector, or what moves the moon round. We have merely,

as I said, assigned to the first the more specific and definite name *gravity*, whereas to the second we assign the more general term impressed power (vita impressa), and the last we call an *intelligence*, either assisting or informing, and as the cause of infinite other motions we give *nature*.

The third day is concerned with the yearly motion of the earth around the sun, presenting arguments for and against Copernicanism. Galileo has Salviati praise Aristarchus and Copernicus for proposing a theory that they arrived at through reasoning, though it seems to be contrary to sensory evidence.

Nor can I sufficiently admire the eminence of those men's intelligence who have received and held it [the heliocentric theory] to be true, and with the sprightliness of their judgments have done such violence to their own senses, that they have been able to prefer that which their reason dictated to them to that which sensible experiences represented to the contrary.… I cannot find any bounds for my admiration how reason was able, in *Aristarchus* and *Copernicus*, to commit such a rape upon their senses as in spite of them, to make herself mistress of her belief.

Galileo describes several celestial phenomena that seem to support the Copernican theory "until it might seem that this must triumph absolutely," but he goes on to say that these arguments will only simplify astronomy and will not show "any necessity imposed by nature." Here, in comparing the two world systems, Galileo is often unfair in his criticism and exaggerates his claims for the superiority of the heliocentric theory.

The fourth day is devoted to Galileo's erroneous theory of tidal action, which he attributed to the threefold Copernican motions of the earth, and which he believed to be conclusive proof of the earth's rotation.

Despite these defects, the arguments for Copernicanism were very persuasive and poor Simplicio, the Aristotelian, is defeated at every turn.

Simplicio's closing remark represents Galileo's attempt to reserve judgment in the debate, where he says that "it would still be excessive boldness for anyone to limit and restrict the Divine power and wisdom to some particular fancy of his own." This statement apparently was almost a direct quote of what Pope Urban had said to Galileo in 1623. When Urban read the *Dialogue* he remembered these words and was deeply offended, feeling that Galileo had made a fool of him and taken advantage of their friendship to violate the 1616 edict against teaching Copernicanism. The Florentine ambassador Francesco Niccolini reported that after discussing the *Dialogue* with Urban, the pope broke out in great anger and fairly shouted, "Your Galileo has ventured to meddle with things that he ought not, and with the most grave and dangerous subjects that can be stirred up these days."

Urban directed the Holy Office to consider the affair and summoned Galileo to Rome. Galileo arrived in Rome in February 1633, but his trial before the court of the Inquisition did not begin until April. There he was accused of having ignored the 1616 edict of the Holy Office not to teach Copernicanism. The court deliberated until June before giving its verdict, and in the interim Galileo was confined in the palace of the Florentine ambassador. He was then brought once again to the Holy Office, where he was persuaded to acknowledge that he had gone too far in his support of the Copernican "heresy," which he now abjured. His abjuration was according to the prescribed formula: "I, Galileo, son of the late Vicenzo Galilei of Florence, seventy years of age ... abandon completely the false opinion that the sun is at the center of the world and does not move and that the earth is not the center of the world and moves."

Galileo was thereupon sentenced to indefinite imprisonment and his *Dialogue* placed on the Index. The sentence of imprisonment was immediately commuted to allow him to be confined in one of the Roman residences of the Medici family, after which he was moved to Siena and then, in April 1634, allowed to return to his villa at Arcetri.

After he returned home Galileo took up again the research he had abandoned a quarter of a century earlier, principally the study of motion. This gave rise to the last and greatest of his works, *Discourses and Mechanical*

Demonstrations Concerning Two New Sciences, of Mechanics and of Motions, which he dictated to his disciple Vincenzo Viviani. The work was completed in 1636, when Galileo was seventy-two and suffering from failing eyesight. Since publication in Italy was out of the question because of the papal ban on Galileo's works, his manuscript was smuggled to Leyden, where the *Discourses* was published in 1638, by which time he was completely blind.

The *Discourses* is organized in the same manner as the *Dialogues*, divided into four days of discussions among three friends. The first day is devoted to subjects that Galileo had not resolved to his satisfaction, particularly his speculations on the atomic theory of matter. The second day was taken up with one of the two new sciences, now known in mechanical engineering studies as strength of materials.

The third and fourth days were devoted to the second of the two new sciences, kinematics, the mathematical description of motion, including motion at constant velocity, uniformly accelerated motion as in free fall, nonuniformly accelerated motion as in the oscillation of a pendulum, and two-dimensional motion as in the path of a projectile. Whereas medieval Latin scholars had discussed all possible kinds of motion, Galileo focused on motions actually found in nature, as he says, "For anyone may invent an arbitrary type of motion and discuss its properties.... But we have decided to consider the phenomena of bodies falling with an acceleration such as actually occur in nature and to make this definition of accelerated motion exhibit the essential features of observed accelerated motion."

The section "On Local Motion," discussed on the third day, presents Galileo's definition of uniformly accelerated motion as a motion in which, "when starting from rest, acquires during equal time intervals equal increments of velocity." He says he adopted this because Nature employs "only those means which are most common, simple and easy."

This definition is equivalent to the modern equation $v = at$, where v is the velocity, a is the constant acceleration, and t is the time. If a is, for example, 2 feet per second, then after 1 second $v = 2$, after 2 seconds $v = 4$, after 3 seconds $v = 6$, etc., all in units of feet per second. Using graphical analysis, derived from Oresme, Galileo showed that s, the distance traversed, increases each second as 1, 3, 5 and so on, which is equivalent to the modern equation $s = \frac{1}{2}at^2$. These two equations, $v = at$ and $s = \frac{1}{2}at^2$, are the basic laws of modern kinematics for the case of uniformly accelerated motion starting from rest. This is the first statement of the laws of kinematics, the purely mathematical description of motion, independent of any dynamical theory. As Salviati says regarding Galileo's purpose in formulating these laws:

> At present it is the purpose of our Author merely to investigate and
> to demonstrate some of the properties of accelerated motion (what-
> ever the cause of the acceleration may be—meaning that the mo-
> mentum of its velocity goes on increasing after proportion from rest
> in simple proportionality to time, which is the same as saying that
> in equal time intervals the body receives equal increments of veloc-
> ity; and if we find that the properties of [accelerated motion] which
> will be demonstrated are realized in freely falling and accelerated
> bodies, we may conclude that the assumed definition includes such
> a motion of heavy bodies and that their speed goes on increasing
> as the time and the duration of the motion.

On the third day Galileo showed that the path of a projectile was the combination of two motions, one of them at constant velocity in the horizontal direction and the other with constant acceleration vertically downward, the resultant of which was a parabola. He also showed that the range of a projectile on a horizontal plane was greatest when the angle of elevation was 45°, and that for angles smaller and larger than this by a given amount the ranges will be equal to one another. Salviati remarks that this showed the superiority of a theoretician able to predict previously unnoticed results, as compared to a purely empirical approach:

> The knowledge of a single fact acquired through the discovery of
> its causes prepares the mind to ascertain and understand other facts
> without need of recourse to experiment, precisely as in the present

case, where by argument alone the Author proves with certainty that the maximum range occurs when the elevation is 45°. He thus demonstrates what perhaps has never been observed in experience, namely, that of other shots those which exceed or fall short of 45° by equal amounts have equal ranges.

Galileo died at Arcetri on January 8, 1642, thirty-eight days before what would have been his seventy-eighth birthday. The Grand Duke of Tuscany sought to erect a monument in his memory, but he was advised not to do so for fear of giving offense to the Holy Office, since the pope had said that Galileo "had altogether given rise to the greatest scandal throughout Christendom."

After Galileo's death a note in his hand was found on the preliminary leaves of his own copy of the *Dialogue*, which he probably wrote after the Holy Office imprisoned him for supporting the Copernican "heresy."

Take note theologians, that in your desire to make matters of faith and of proposition relating to the fixity of sun and earth you may run the risk of eventually having to condemn as heretics those who would decide the earth to stand still and the sun to change position—eventually, I say—at such a time as it might be physically or logically proved that the earth moves and the sun stands still.

14

On the Shoulders of Giants

*T*HE PIONEERING OBSERVATIONS AND NEW THEORIES OF COPERNI-
cus, Tycho Brahe, Kepler, and Galileo, together with those of
some of their contemporaries, represent part of an intellectual up-
heaval that came to be called the Scientific Revolution, which continued
through the seventeenth century and on into the early years of the eighteenth, a period during which the worldview of western Europe changed
profoundly and modern scientific culture emerged.

Some historians today take issue with the concept of a Scientific Revolution, as historian and sociologist Steven Shapin remarked in his definitive
work on the subject:

> Many historians are now no longer satisfied that there was any sin-
> gular and discrete event, localized in time and space, that can be
> pointed to as "the" Scientific Revolution. Such historians now reject
> even the notion that there was even any single coherent entity called
> "science" in the seventeenth century to undergo revolutionary
> change. There was, rather, a rather diverse array of cultural prac-
> tices aimed at understanding, explaining and controlling the natural
> world, each with different characteristics and each experiencing dif-
> ferent modes of change. We are now much more dubious of claims
> that there is anything like a "scientific method"—a cultural, uni-
> versal, and efficacious set of procedure for making scientific knowl-
> edge—and still more skeptical of stories that locate its origins in

the seventeenth century, from which time it has been unproblematically passed on to us.

One particularly influential system of natural philosophy that emerged in the seventeenth century was mechanism, which held that all natural phenomena was due to one single kind of change, the motion of matter. The approach of the mechanistic philosophy of nature is broadly summarized by Pierre Gassendi (1592–1565), a French Catholic priest who in 1647 published a work in which he attempted to reconcile the atomic theory with Christian doctrine. "There is no effect without a cause; no cause acts without motion, nothing acts on distant things except through itself or an organ or transmission; nothing moves unless it is touched, whether directly or through an organ or through another body."

Gassendi's mechanism was based on the atomic theory as interpreted by Lucretius, in which physical properties are traced to the imagined size and shape of the component particles. Gassendi was deeply influenced by Isaac Beeckman (1566–1637), who expressed his corpuscular form of mechanism in the statement that "all properties arise from [the] motion, shape and size [of the fundamental particles]. So that each of these three things must be considered."

A number of different approaches to scientific investigation were formulated in the seventeenth century. One was the empirical, inductive method proposed by Francis Bacon (1561–1626); another was the theoretical, deductive approach of René Descartes (1596–1650).

According to Bacon, the new science should be based primarily on observation and experiment, and it should arrive at general laws only after a careful and thorough study of nature. In his *Novum organum*, published in 1620, Bacon criticized the existing state of scientific knowledge. "The subtlety of nature greatly exceeds that of sense and understanding, so that those fine meditations, speculations and fabrications of mankind are unsound, but there is no one to stand by and point it out. And just as the sciences we now have are useless for making discoveries of practical use, so the present logic is useless for the discovery of the sciences."

Bacon never accepted the Copernican theory, which he called a hypothesis, and he criticized both Ptolemy and Copernicus for presenting nothing more than "calculations and predictions" rather than "philosophy … what is found in nature herself, and is actually and really true."

Descartes sought to give physical laws the same certitude as those of mathematics. As he wrote in a letter to Marin Mersenne: "In physics I should consider that I knew nothing if I were able to explain only how things might be, without demonstrating that they could not be otherwise. For having reduced physics to mathematics, this is something possible, and I think that I can do it within the small compass of my knowledge, though I have not done it in my essays."

Descartes writes of how he proposed to himself four "laws of reasoning," which he applied first to the study of mathematics:

> In this way I believed I could borrow all that was best both in geometrical analysis and in algebra, and correct all the defects of the one by the help of the other. And, in points of fact, the accurate observance of these few precepts gave me such ease in unraveling all the questions embraced in these two sciences, that in the two or three months I devoted to their examination, not only did I reach solutions of questions I had formerly deemed exceedingly difficult, but even as regards questions of the solutions of which I remained ignorant, I was enabled as it appeared to me, to determine the means whereby, and the extent to which, a solution was possible.

Whereas in philosophy Descartes began with the existence of self (*Cogito ergo sum*, I am thinking, therefore I exist), in physics he started with the existence of matter, its extension in space, and its motion through space. That is, everything in nature can be reduced to matter in motion. Matter exists in discrete particles that collide with one another in their ceaseless motions, changing their individual velocities in the process, but with the total "quantity of motion" in the universe remaining constant. Descartes wrote of the divine origin of this law in his *Principles of Philosophy* (1644), an extraordinarily detailed and elaborate model of the physical universe based on his corpuscular mechanistic theory. Speaking of God, he said, "In the beginning, in his omnipotence, he created matter, along with its motion and rest, and now, merely by his regular concurrence, he preserves the same amount of motion and rest in the material universe as he put there in the beginning."

Descartes presented his method in *Rules for the Direction of the Mind*, completed in 1628 but not published until after his death, and in the *Discourse on Method*, published in 1637 along with appendices entitled *Optics, Geometry*, and *Meteorology*. He gave the final form of his three laws of nature in *The Principles of Philosophy* (1644). The first law, the principle of inertia, states that "each and every thing, insofar as it can, always continues in the same state, and thus what is once in motion always continues to move." The second law, dealing with directionality, states that "all motion is in itself rectilinear ... every piece of matter, considered in itself, always tends to continue moving, not in any oblique path but only in a straight line." The third law is concerned with collisions: "If a body collides with another body that is stronger than itself, it loses none of its motion; but if it collides with a weaker body, it loses a quantity of motion equal to that which it imparts to the other body."

The *Optics* presents Descartes' mechanistic theory of light, which he conceived of as a series of impulses propagated through the finely dispersed microparticles that fill the spaces between macroscopic bodies, leaving no intervening vacuum. This model gave him the right form for the law of refraction, but in his derivation he took the velocity of light to be greater in water than in air, which is not true.

The *Geometry* was inspired by what Descartes called the "true mathematics" of the ancient Greeks, particularly Pappus and Diophantus. Here he provided a geometric basis for algebraic operations, which to some extent had already been done by his predecessors as far back as al-Khwarizmi. The symbolic notation used by Descartes quickly produced great progress in algebra and other branches of mathematic. His work gave rise to the branch of mathematics now known as analytic geometry, which had been anticipated by Pierre Fermat (1539–1565). Fermat, inspired by Diophantus and Apollonius, was also one of the founders of modern number theory and probability theory.

Descartes' *Meteorology* includes his model of the rainbow, in which he used the laws of reflection and refraction to obtain the correct values of the angles at which the primary and secondary bows appear. He begins his explanation by pointing out that rainbows occur not only in the sky but also in illuminated fountains and sprays, so that it is not solely a celestial phenomenon but rather one involving light and individual drops of water. He tested this hypothesis by taking a spherical glass flask full of water, holding it up at arm's length in the sunlight, and moving it up and down so that colors are produced. He says that if he stood with his back to the sun so as to let the light come

from the part of the sky marked AFZ and my eye be at E, then when I put this sphere at the place BCD the part of it at D seems to me wholly red and incomparably more brilliant than the red. And whether I move toward it or step back from it, and move to the right or the left, or even turn it in a circle around my head, then provided the line DE always makes an angle of around 42° with the line EM, which one must imagine to extend from the center of the eye to the center of the sun, D always appears equally red. But as soon as I made the angle DEM the slightest bit larger, the redness disappeared. And when I made it a little bit smaller it did not disappear completely in one stroke but first divided as into two less brilliant parts in which could be seen yellow, blue and other colours. Then, looking towards the place marked K on the sphere, I perceived that, making the angle KEM around 52°, K also appeared to be colored red, but not as brilliant as D.

Then, using a sheet of black paper to screen off selected parts of the sunlight, Descartes determined the path of the rays producing the various colors of both the primary and secondary rainbows. He con-

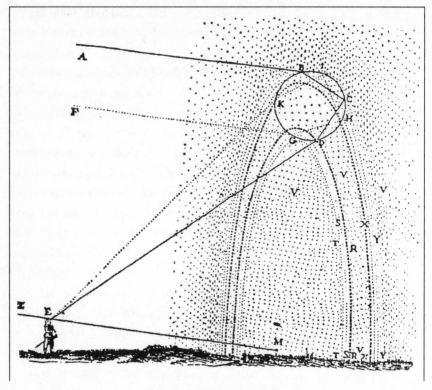

Descartes' model of the rainbow

cluded, as had Dietrich of Freiburg, that the primary rainbow was produced by two refractions and one internal reflection within each of the
raindrops, while the secondary bow was generated by two refractions
and two internal reflections, the second of which had the effect of inverting the spectrum.

Descartes then performed several experiments to show that the actual
dispersion of light into colors was due only to refraction and not reflection.
In one of these experiments he dispersed sunlight in a glass prism and observed the colors on a screen. When sunlight strikes the face perpendicularly, it passes through the glass to a narrow aperture in the darkened base,
where it is refracted and dispersed into a band HGF on the screen, with violet appearing above, red below, and the other colors in between. He then

tried to explain why the dispersion takes place and "why these colors are different above and below, even though the refraction, shadow and light concur there in the same way." The explanation that he proposed was an extremely detailed mechanistic model based on his microcorpuscular theory of matter. One of the assumptions that Descartes had made in this theory is that light travels more rapidly in dense media such as water and glass than in air, which is incorrect.

Chapters 8 through 12 of Descartes' *Le monde* present his mechanistic cosmology, based on his theory of matter and laws of motion. This hypothetical "new world" that he described consisted of an indefinite number of contiguous vortices, each with a star at its center. He argued that the stars were the sources of light just like our sun, for "if we consider how bright and glittering the rays of the fixed stars are, despite the fact that they are an immense distance from the sun, we will not find it hard to accept that they are not like the sun. Thus if we are as close to one of them as we are to the sun, that star would in all probability appear as large and luminous as the sun." He held that each of these stars is the center of a planetary system, all carried around by the motion of the particles of the three types of matter that he believed filled all of space.

Descartes' vortex theory was generally accepted at first, but the research of Christiaan Huygens (1629–1695) showed conclusively that it was completely incorrect. Huygens was led to his rejection of the vortex theory by his studies of dynamics. In one of his studies he considered a situation in which a lead ball is attached to a string held by a man standing at the center of a rotating platform. When the platform rotates the man feels an outward or centrifugal force in the string attached to the ball, which in turn experiences an inward or centripetal force due to the string. Huygens found that the centripetal force on the ball was directly proportional to the mass of the ball and the square of its velocity, and inversely proportional to the radius of its circular path, thus establishing the basis of dynamics for circular motion. This and his researches on the laws of collisions were what led Huygens to conclude that the Cartesian cosmology was in error. As he said in 1693, he could find "almost nothing I can approve as true in all the physics and metaphysics" of Descartes.

Huygens found that Descartes' rules concerning collisions were not mutually consistent. Descartes held there was both relativity of motion and conservation of motion, while Huygens realized that these are incompatible with each other. He saw that his task was to clarify just what relativity of motion implied for collisions.

In his *De motu corporum ex percussione*, the first version of which was completed in the mid-1550s, Huygens framed an ingenious thought experiment, carried out by two men, one of them in a boat moving at constant speed along a canal and the other standing on the shore. The man in the boat has two identical metal spheres attached by strings from his outstretched hands, which he brings together to make them collide elastically just as he passes the man on the shore, who joins hands with him at that moment, so that they are in effect doing the experiment together. If the boat is moving at the same speed as one of the bodies, then the man on the shore will see a moving body colliding with one at rest, the latter being given a velocity equal to that which the former had before collision. The man on the boat, on the other hand, sees the two bodies as simply interchanging their velocities. All other cases involving the collisions of two identical bodies can be accommodated to this symmetrical case simply by changing the speed of the boat.

By looking at the same phenomenon in two different frames of reference, Huygens showed that the center of gravity of a system is unchanged in an elastic collision. That is, if the center of gravity is at rest, as it is for the man on the shore, then it will remain at rest despite the collision, whereas if it is moving with constant speed, as it is for the man in the boat, then it will continue to do so.

Huygens also did pioneering work on motion in his treatise on the pendulum clock, the *Horologium oscillilatorium*, published in 1673. The thesis describes an isochronus pendulum clock invented by Huygens in 1656, in which the pendulum bob swings against a cycloidal surface, which makes its period independent of amplitude for all angles. Huygens describes

Huygens' demonstration of the relativity of motion

the significance of his research, noting that "the simple pendulum does not naturally provide an accurate and equal measure of time since the wider motions are observed to be slower than the narrower ones. But by a geometrical method we have found a different and previously unknown way to suspend the pendulum; and we have discovered a line whose curvature is marvelously and quite rationally suited to give the required equality to the pendulum."

Part 2 of the *Horologium* begins with three hypotheses on dynamics. The first, which is a clear statement of the principle of inertia, states that in the absence of gravity a body will continue in any motion it already has in a straight line at constant velocity. The second hypothesis is that gravity always acts so as to impose a downward component on any uniform motion the body has, and the third says that these motions are independent of one another.

Descartes' Aristotelian notion that a vacuum was impossible was also shown to be incorrect by several of his contemporaries, beginning with Evangelista Torricelli (1608–1647) and Blaise Pascal (1623–1662). Torricelli's invention of the barometer in 1643 led him to conclude that the closed space above the mercury column represented at least a partial vacuum, and that the difference in the height of the two columns in the U-tube was a measure of the weight of a column of air extending to the top of the atmosphere. Pascal had a barometer taken to the top of the Puy de Dôme, a peak in central France, and it was observed that the difference in height of the two columns was less than at sea level, verifying Torricelli's conclusions. The results of this experiment led Pascal to urge all disciples of Aristotle to see if the writings of their master could explain the results. "Otherwise," he wrote, "let them recognize that experiments are the real masters that we should follow in physics; that the experiment done in the mountains overturns the universal belief that nature abhors a vacuum."

The German engineer Otto von Guericke (1602–1680) discovered that it was possible to pump air as if it were water, allowing him to produce a vacuum mechanically. In a famous experiment in Magdeburg in 1657, he pumped the air out of a spherical cavity made by fitting together two copper hemispheres and showed that the resulting differential pressure was so great that not even two teams of horses pulling in opposite directions could force the two halves of the sphere apart.

Guericke's demonstration led the Irish chemist Robert Boyle (1627–1691) to have a vacuum pump constructed by the instrument maker Ralph Greatorex. The design of the pump was subsequently improved by the English physicist Robert Hooke (1635–1703). Boyle fitted the pump with a Torricellian barometer and noted the change in the level of mercury as the tube was evacuated. He then used the pump to do research on pneumatics, which he published in 1660 under the title *New Experiments Physico-Mechanicall, Touching the Spring of Air and Its Effects*. He concluded that a vacuum can be produced, or at least a partial one; that sound does not propagate in a vacuum; and that air is necessary for life or a flame. Additionally, air is an elastic fluid that exerts a pressure against whatever restricts it and expands when it is relieved of external constraints: "Air ether consists of, or at least abounds with, parts of such a nature, that in case they be bent or compress'd by the incumbent part of thermosphere, they do endeavour, as much as in them lies, to free themselves from that pressure, by bearing upon the contiguous bodies that keep them bent." In an appendix to the second edition of this work, published in 1662, he established the relationship now known as Boyle's law, that the pressure exerted by a gas is inversely proportional to its volume.

Boyle was influenced by both Francis Bacon's empiricism and Descartes' mechanistic view of nature. He was also influenced by the natural philosophy of Epicurus, revived by Pierre Gassendi. This led Boyle to adopt a divinely ordained corpuscular version of mechanism, which he described in his treatise on *Some Thoughts About the Excellence and Grounds of the Mechanical Philosophy*, published in 1674. As he concluded concerning the universality of mechanism: "By this very thing that the mechanical principles are so universal, and therefore applicable to so many other things, they are rather fitted to include, than necessitated to exclude, any other hypothesis, that is founded in nature, as far as it is so."

The scientific developments that had taken place from the time of Copernicus through that of Galileo culminated with the career of Isaac Newton (1642–1727), whose supreme genius made him the central figure in the emergence of modern science.

Newton was born on December 25, 1642, in the same year that Galileo had died. His birthplace was the manor house of Woolsthorpe in Lincolnshire, England. His father, an illiterate farmer, had died three months before Isaac was born, and his mother remarried three years later, though she was widowed again after eight years. When Newton was twelve he was enrolled in the grammar school at the nearby village of Grantham, and he studied there until he was eighteen. His maternal uncle, a Cambridge graduate, sensed that his nephew was gifted and persuaded Isaac's mother to send the boy to Cambridge, where he was enrolled at Trinity College in June 1661.

At Cambridge Newton was introduced to Aristotelian science and cosmology as well as the new physics, astronomy, and mathematics of Copernicus, Kepler, Galileo, Fermat, Descartes, Huygens, and Boyle. In 1663 he began studying under Isaac Barrow (1630–1677), the newly

Isaac Newton, 1702

appointed Lucasian professor of mathematics and natural philosophy. Barrow edited the works of Euclid, Archimedes, and Apollonius, and published his own works on geometry and optics, with the assistance of Newton.

By Newton's own testimony, he began his researches in mathematics and physics late in 1664, shortly before an outbreak of plague closed the university at Cambridge and forced him to return home. During the next two years, his anni mirabilis, he began his research in calculus and the dispersion of light, and discovered his law of universal gravitation and motion as well as the concepts of centripetal force and acceleration.

> In the beginning of the year 1665 I found the Method of approximating series & the Rule for reducing any dignity of any Binomial

into such a series. The same year in May I found the method of Tangents of Gregory & Slusius, and in November had the direct method of fluxions & the next year in January had the Theory of Colours & in May following I had entrance into ye inverse method of fluxions. And the same year I began to think of gravity extending to ye orb of the Moon & (having found out how to estimate the force with which [a] globe revolving within a sphere presses the surface of the sphere) from Keplers rule of the periodic times of the Planets being sesquialternate proportion of their distances from the center of their Orbs. I deduced that the forces which keep the Planets in their Orbs must [be] reciprocally as the squares of their distances from the centers about which they revolve: and thereby compared the force required to keep the Moon in her Orb with the force of gravity at the surface of the earth & found them answer pretty nearly. All this was in the two plague years 1665 and 1666 for in those years I was in the prime of my age for invention & minded Mathematicks & Philosophy more than at any time since.

This indicates that Newton had derived the law for centripetal force and acceleration by 1666, some seven years before Huygens, though he did not publish it at the time. He applied the law to compute the centripetal acceleration at the earth's surface caused by its diurnal rotation, finding that it was less than the acceleration due to gravity by a factor of 250, thus settling the old question of why objects are not flung off the planet by its rotation. He computed the centripetal force necessary to keep the moon in orbit, comparing it to the acceleration due to gravity at the earth's surface, and found that they were inversely proportional to the squares of their distances from the center of the earth. Then, using Kepler's third law of planetary motion together with the law of centripetal acceleration, he verified the inverse square law of gravitation for the solar system. At the same time he laid the foundations for calculus and formulated his theory for the dispersion of white light into its component colors.

In the spring of 1667, when the plague subsided, Newton returned to

Cambridge. Two years later he succeeded Barrow as Lucasian professor of mathematics and natural philosophy, a position he was to hold for nearly thirty years.

During the first few years after he took up his professorship, Newton devoted much of his time to research in optics and mathematics. He continued his experiments on light, examining its refraction in prisms and thin glass plates as well as working out the details of his theory of colors. He built a reflecting telescope with a magnifying power of nearly forty, and then made a refractor that he claimed magnified 150 times, using it to observe planets and comets. The latter telescope came to the attention of the Royal Society, which elected him a Fellow on January 11, 1672.

As part of his obligations as a Fellow, Newton wrote a paper on his optical experiments, which he submitted on February 28, 1672, to be read at a meeting of the Society. The paper, subsequently published in the *Philosophical Transactions of the Royal Society*, described his discovery that sunlight is composed of a continuous spectrum of colors, which can be dispersed by passing light through a refracting medium such as a glass prism. He found that the "rays which make blue are refracted more than the red," and he concluded that sunlight is a mixture of light rays, some of which are refracted more than others. Furthermore, once sunlight is dispersed into its component colors it cannot be further decomposed. This meant that the colors seen on refraction are inherent in the light itself and are not imparted to it by the refracting medium.

The paper was characteristic of Newton's attitude toward the approach to be followed in any scientific investigation. Later, in a controversy arising out of his first paper, Newton described his scientific method.

> For the best and safest method of philosophizing seems to be, first to enquire diligently into the properties of things, and to establish these properties by experiment, and then to proceed more slowly to hypotheses for the explanation of them. For hypotheses should be employed only in explaining the properties of things, but not assumed in determining them, unless so far as they may furnish experiments.

Ironically, the paper was widely criticized by Newton's contemporaries for just the contrary reason: that it did not confirm or deny any general philosophy of nature, and the mechanists objected that it was impossible to explain his findings on the basis of any mechanical principles. Then there were others who insisted that Newton's experimental findings were false, since they themselves could not find the phenomena that he had reported. Newton replied patiently to each of these criticisms in turn, but after a time he began to regret ever having presented his work in public.

One of those who criticized his paper was Robert Hooke, who in November 1662 been had been appointed as the first curator of experiments at the newly founded Royal Society, a position he held until his death in 1703, making many important discoveries in mechanics, optics, astronomy, technology, chemistry, and geology. His lengthy critique of the paper seemed to imply that Hooke had performed all of Newton's experiments himself, while rejecting the conclusions that Newton had drawn.

This led Newton to resign from the Royal Society early in 1673, but Henry Oldenburg, secretary of the Society, refused to accept his resignation and persuaded him to remain. Then in 1676, after a public attack by Hooke, Newton broke off almost all association with the Oldenburg and the Royal Society. The following year Oldenburg died and Hooke replaced him as secretary, whereupon he wrote a conciliatory letter in which he expressed his admiration for Newton. Referring to Newton's theory of colors, Hooke said that he was "extremely well pleased to see those notions promoted and improved which I long since began, but had not time to compleat."

Newton replied in an equally conciliatory tone, referring to Descartes' work on optics. "What Descartes did was a good step. You have added much several ways, and especially in taking the colors of thin plates into philosophical consideration."

But despite these friendly sentiments, the two were never completely reconciled, and Newton maintained his silence. Nevertheless they continued to communicate with each other, a correspondence that was to lead again and again to controversy, the bitterest dispute arising from Hooke's claim that he had discovered the law of gravitation before Newton.

By 1684 others beside Hooke and Newton were convinced that the gravitational force was responsible for holding the planets in their orbits and that this force varied with the inverse square of their distance from the sun. Among them were the astronomer Edmund Halley (1656–1742), a good friend of Newton's and a fellow member of the Royal Society. Halley made a special trip to Cambridge in August 1684 to ask Newton what he thought the curve would be that would be described by the planets supposing the force of attraction toward the sun to be reciprocal to the square of their distance from it. Newton replied immediately that it would be an ellipse, but he could not find the calculation, which he had done seven or eight years before. And so he was forced to rework the problem, after which he sent the solution to Halley that November.

By then Newton's interest in the problem had revived, and he developed enough material to give a course of nine lectures in the fall term at Cambridge, under the title of *De motu corporum* (*The Motion of Bodies*). When Halley read the manuscript of *De motu* he realized its immense importance, and he obtained Newton's promise to send it to the Royal Society for publication. Newton began preparing the manuscript for publication in December 1684 and sent the first book of the work to the Royal Society on April 28, 1686.

On May 22, Halley wrote to Newton saying that the Society had entrusted him with the responsibility for having the manuscript printed. But he added that Hooke, having read the manuscript, claimed that it was he who had discovered the inverse square nature of the gravitational force and thought that Newton should acknowledge this in the preface. Newton was very much disturbed by this, and in his reply to Halley he went to great lengths to show that he had discovered the inverse square law of gravitation and that Hooke had not contributed anything of consequence.

The first edition of Newton's work was published in midsummer 1687 at the expense of Halley, since the Royal Society had found itself financially unable to fund it. Newton entitled his work *Philosophicae naturalis principia mathematica* (*The Mathematical Principles of Natural Philosophy*), referred to more simply as the *Principia*. The *Principia* begins with an ode

dedicated to Newton by Halley. This is followed by a preface in which Newton outlines the scope and philosophy of his work.

> Our present work sets forth mathematical principles of natural
> philosophy. For the basic problem of philosophy seems to be to
> discover the forces of nature from the phenomena of motions,
> and then to demonstrate the other phenomena from these
> forces.... Then the motions of the planets, the comets, the moon,
> and the sea are deduced from these forces by propositions that are
> also mathematical. If only we could derive the other phenomena of
> nature from mechanical principles by the same kind of reasoning!

Book 1 begins with a series of eight definitions, of which the first five
are fundamental to Newtonian dynamics. The first effectively defines
"quantity of matter," or mass, as being proportional to the weight density
times volume. The second defines "quantity of motion," subsequently to be
called "momentum," as mass times velocity. In the third definition Newton
says that the "inherent force of matter," or inertia, "is the power of resisting
by which every body, so far as it is able, perseveres in its state either of rest
or of moving uniformly straight forward." The fourth states that "impressed
force is the action exerted upon a body to change its state either of resting
or of uniformly moving straight forward." The fifth through eighth define
centripetal force as that by which bodies "are impelled, or in any way tend,
toward some point as to a center." As an example Newton gives the gravitational force of the sun, which keeps the planets in orbit.

As regards the gravity of the earth, he gives the example of a lead
ball, projected from the top of a mountain with a given velocity, and in a
direction parallel to the horizon. If the initial velocity is made larger and
larger, he says, the ball will go farther and farther before it hits the ground,
and may go into orbit around the earth or even escape into outer space.

The definitions are followed by a scholium, a lengthy comment in
which Newton gives his notions of absolute and relative time, space, place,
and motion. These essentially define the classical laws of relativity, which

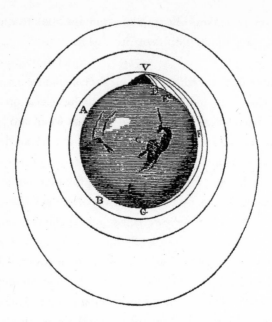

An illustration from Newton's Principia, *showing a projectile in orbit around the Earth.*

in the early twentieth century would be superseded by Einstein's theories of special and general relativity.

Next come the axioms, now known as Newton's laws of motion, three in number, each accompanied by an explanation and followed by corollaries.

> Law 1: Every body perseveres in its state of being at rest, or of moving uniformly forward, except insofar as it is compelled to change its state of motion by forces impressed.
>
> Law 2: A change of motion is proportional to the motive force impressed and takes place along the straight line in which that force is impressed.
>
> Law 3: To every action there is always an opposite and equal reaction; in other words, the action of two bodies upon each other are always equal, and always opposite in direction.

The first law is the principal of inertia, which is actually a special case of the second law when the net force is zero. The form used today for the second law is that the force F acting on a body is equal to the time rate of change of the momentum p, where p equals the mass m times the velocity v; if the mass is constant then $F = ma$, a being the acceleration, the time rate of change of the velocity. The third law says that when two bodies interact, the forces they exert on each other are equal in magnitude and opposite in direction.

The introductory section of the *Principia* is followed by book 1, entitled "The Motion of Bodies." This begins with an analysis of motion in general, essentially using calculus. First Newton analyzed the relations between orbits and central forces of various kinds. From this he was able to show that if and only if the force of attraction varies as the inverse square of the distance from the center of force, then the orbit is an ellipse, with the center of attraction at one focal point, thus proving Kepler's second law of motion. Elsewhere in book 1 he proves Kepler's first and third laws. He also uses his third law of motion to deal with problems involving two bodies mutually attracting each other, where he notes that neither of the two bodies can be considered to be at rest: "For attractions are always directed toward bodies, and—by the third law—the actions of attracting and attracted bodies are always mutual and equal; so that if there are two bodies, neither the attracting and attracted body can be at rest, but both … revolve around a common center of gravity as if by mutual attraction."

Book 2 is also entitled "The Motion of Bodies," for the most part dealing with forces of resistance to motion in various types of fluids. One of Newton's purposes in this analysis was to see what effect the hypothetical aether in Descartes' cosmology would have on the motion of the planets. His studies showed that the Cartesian vortex theory was completely erroneous, for it ran counter to the laws of motion in resisting media that he established in book 2 of the *Principia*.

The third and final book of the *Principia* is entitled "The System of the World," beginning with three "Rules for the Study of Natural Philosophy." After this comes a section on "Phenomena," six in number, followed by forty-two propositions, each accompanied by a theorem and lemmas, sometimes followed by a scholium. This is in turn followed by a general scholium and a concluding section entitled "The System of the World."

The six phenomena concern the motion of the planets and the earth's moon, along with observations concerning Kepler's second and third laws of planetary motion. He concludes that the planets, "by radii drawn to the center ..., describe areas proportional to the times, and their periodic times—the fixed stars being at rest—are as the 3/2 powers of their distances from that center."

The various propositions and lemmas are concerned with the consequences of Newton's theory for both terrestrial and celestial motion. His law of universal gravitation states that "gravity exists in all bodies universally and is proportional to the quantity of matter in each." It also states that the gravitational force between two bodies varies as the inverse square of the distance between their centers of mass. This inverse-square gravitational force explains the motion of the planets orbiting the sun, the satellites of Jupiter, and the earth's moon, as well as the local gravity on the earth itself. Proposition 13 states Kepler's first and second laws of planetary motion: "The planets move in ellipses that have a focus in the center of the sun, and by radii drawn to that center they describe areas proportional to the times."

Telescopic observation of the planets had revealed that they were oblate spheres, as stated in proposition 18: "The axes of the planets are smaller than the diameters that are drawn perpendicular to the axes," that is, the planets are oblate spheres. Newton correctly attributed this effect to the centrifugal forces arising from the axial rotation of the planets, so that the earth, for example, is flattened at the poles and bulges around the equator.

Newton's theory of tidal action, is stated in proposition 24: "The ebb and flow of the sea arises from the actions of the sun and moon," finally solving a problem that dated back to the time of Aristotle. Another ancient problem is stated in proposition 39: "Find the precession of the equinoxes," including the gravitational forces of both the sun and the moon on the earth. Newton correctly computed that "the precession of the equinoxes is more or less 50 seconds [of arc] annually," thus solving another problem that had preoccupied astronomers for some two thousand years.

Still another ancient unsolved question is settled in lemma 4, which states that "the comets are higher than the moon, and move in the planetary regions." In the lemmas and propositions that follow, Newton discusses the motion of comets, showing that they move in elliptical orbits around the sun, thus reappearing periodically, as did the one known as Halley's comet, which had been observed in 1682 after disappearing seventy-five years before. He also speculated on the nature of comets, saying, as had Kepler, that the tail of a comet represents vaporization from the comet's head as it approaches the sun.

This is followed by a general scholium, in which Newton says that mechanism alone cannot explain the universe, whose harmonious order indicated to him the design of a Supreme Being. "This most elegant system of the sun, planets, and comets could not have arisen without the design and dominion of an intelligent and powerful being."

A second edition of the *Principia* was published in 1713 and a third in 1726, in both cases with a preface written by Newton. Meanwhile in 1704 Newton had published his research on light, much of which had been done early in his career. Unlike the *Principia*, which was in Latin, the first edition of his new work was in English, entitled *Opticks, or a Treatise of the Reflexions, Refractions, Inflexions and Colours of Light*. The first Latin edition appeared in 1706, and subsequent English editions appeared in 1717/1718, 1721, and 1730; the last, which came out three years after Newton's death, bore a note stating that it was "corrected by the author's own hand, and left before his death, with his bookseller."

Like the *Principia*, the *Opticks* is divided into three books. At the very beginning of book 1 Newton reveals the purpose he had in mind when composing his work. "My design in this Book," he writes, "is not to explain the Properties of Light by Hypotheses, but to propose and prove them by Reason and Experiment."

The topics dealt with in book 1 include the laws of reflection and refraction, the formation of images, and the dispersion of light into its component colors by a glass prism. Other topics include the properties of lenses and Newton's reflecting telescope; the optics of human vision; the theory of the rainbow; and an exhaustive study of color. Newton's proof of the law of refraction is based on the erroneous notion that light travels more rapidly in glass than in air, the same error that Descartes had made. This error stems from the fact that both of them thought that light was corpuscular in nature.

Newton's corpuscular view of light stemmed from his acceptance of the atomic theory. He wrote of his admiration for "the oldest and most celebrated Philosophers of Greece ... who made a Vacuum, and Atoms, and the Gravity of Atoms, the first Principles of their Philosophy.... All these things being consider'd, it seems to me that God in the Beginning formed Matter in solid, hard, impenetrable, moveable Particles, of such Sizes and Figures, and with such other Properties and in such Proportions to Space, as much conduced to the End for which he had form'd them."

Book 2 begins with a section entitled "Observations Concerning the Reflexions, Refractions, and Colours of Thin Transparent Bodies." The effects that he studied here are now known as interference phenomena, where Newton's observations are the first evidence for the wavelike nature of light.

In book 2 Newton also comments on the work of the Danish astronomer Olaus Roemer (1644–1710), who in 1676 measured the velocity of light by observing the time delays in successive eclipses of the Jovian moon Io as Jupiter receded from the earth. Roemer's value for the velocity of light was about a fourth lower than the currently accepted one of slightly less than 300,000 kilometers (186,410 miles) per second, but it was nevertheless the first measurement to give an order of magnitude estimation of one of the fundamental constants of nature. Roemer computed that light would take eleven minutes to travel from the sun to the earth, as compared to the correct value of eight minutes and twenty seconds. Newton seems to have made a better estimate of the speed of light than Roemer, for in book 2 of the *Opticks* he says that "Light is propagated from luminous Bodies in time, and spends about seven or eight Minutes of an Hour in passing from the Sun to the Earth."

In book 3 the opening section deals with Newton's experiments on diffraction. The remainder of the book consists of a number of hypotheses, not only on light, but on a wide variety of topics in physics and philosophy. The first edition of the *Opticks* had sixteen of these queries, the second twenty-three, the third and fourth forty-one each. It would seem that Newton, in the twilight of his career, was bringing out into the open some of his previously undisclosed speculations, his heritage for those who would follow him in the study of nature.

Meanwhile Newton had been involved in a dispute with the great German mathematician and philosopher Gottfried Wilhelm Leibniz (1646–1716), the point of contention being which of them had been the first to develop calculus. According to his own account, Newton first conceived the idea of his "method of fluxions" around 1665–1666, although he did not publish it until 1687, when he used it in the *Principia*. He first published his work on calculus independently in a treatise that came out in 1711. Leibniz began to develop the general methods of calculus in 1675, though he did not publish his work until 1684. The version of calculus formulated by Leibniz, whose notation was much like that used today, caught on more rapidly than that of Newton, particularly on the Continent. Newton's bitterness over the dispute was such that in the third edition of the *Principia* he deleted all reference to Leibniz, who until the end of his days continued to accuse his adversary of plagiarism.

Aside from his work in science, Newton also devoted much of his time to studies in alchemy, prophecy, theology, mythology, chronology, and history. His most important nonscientific work is *Observations upon the Prophecies of Daniel, and the Apocalypse of St. John*, which is considered to be a possible key to the method of his alchemical studies, as evidenced by such notions as his analogy between the "four metals" of alchemy and the four beasts of the apocalypse.

In 1689 Newton was elected by the constituency of Cambridge University to serve as member of Parliament. He was made warden of the mint in March 1696, whereupon he appointed William Wiston as his deputy in the Lucasian professorship at Cambridge. He finally resigned his professorship on March 10, 1710, shortly after his second election as MP for the university. He was knighted by Queen Anne at Trinity College on April 16, 1705; on May 17, 1706, he was defeated in his third campaign for the university's seat in Parliament.

Newton died in London on March 20, 1727, four days after presiding over a meeting of the Royal Society, of which he had been president since 1703. His body lay in state until April 4, when he was buried with great pomp in Westminster Abbey. The baroque monument marking his tomb shows Newton in a reclining position, along with a weeping female figure, representing Astronomy, Queen of the Sciences, sitting on a globe above. The inscription on the tomb concludes with: "Let Mortals rejoice That there has existed such and so great an Ornament to the Human Race."

His contemporaries hailed Newton's achievement as the perfection of the mechanistic philosophy of nature, and historians of the mid-twentieth century praised his work as the culmination of the Scientific Revolution. Although many historians of the new millennium now take issue with the notion of a Scientific Revolution, it is generally agreed that Newton's work culminated the long development of European science, creating a synthesis that opened the way for the scientific culture of the modern age.

Newton himself paid tribute to his predecessors when he said, in response to Hooke, "If I have seen further than Descartes, it is by standing on the sholders [sic] of Giants." Newton was echoing the statement that had made five centuries earlier by Bernard Silvestre.

Bernard was writing just before the founding of the first of the new European universities, opening up Graeco-Islamic science and philosophy to Latin Europe. Thus the predecessors he was referring to were the scholarly monks who had toiled away obscurely in the monasteries of Ireland and England and then on the Continent, studying the scraps of classical learning that had survived the collapse of the Graeco-Roman world and trying to understand the world around them, cultivating the nascent scientific tradition that would come to fruition with the works of Copernicus, Kepler, Galileo, and Newton.

Newton himself was referring not only to his European predecessors, but also to the ancient Greek philosophers and scientists from Thales and Pythagoras and the other pre-Socratics through Democritus, Plato, Aristotle, Euclid, Archimedes, Apollonius, Aristarchus, Eratosthenes, and Ptolemy, whose manuscripts had been lost in the burning of the great Library of Alexandria, the lost knowledge slowly recovered in medieval Byzantium, the Middle East, and Europe until finally, more than twelve centuries later, it gave rise to the Newtonian synthesis that began modern science.

Newton himself was referring not only to his European predecessors but also to the ancient Greek philosophers and scientists from the Presocratics up through the Hellenistic period, whose manuscripts had been destroyed in the burning of the great Library of Alexandria, the lost knowledge slowly recovered and further developed in medieval Byzantium, Islam and Europe, paving the way for the Newtonian synthesis.

Beginning a thousand years before Galileo, with one thinker passing on ideas to another, enlightenment gradually dispelled the darkness and led to the dawning of the modern scientific age.

Bibliography

ABBREVIATIONS USED

DBS   *Dictionary of Scientific Biography.* 16 vols. Edited by Charles Coulston Gillispie. New York, 1970–1980.

Tradition, Transmission, Transformation
   Ragep, F. Jamil, and Sally P. Ragep with Steven Livesey (eds.). *Tradition, Transmission, Transformation.* Leiden, 1996.

Africa, Thomas W. "Copernicus' Relation to Aristarchus and Pythagoras," *Isis* 52, no. 3 (September 1961): 403–9.

Ahmad, S. Maqbal. "Al-Idisi," *DSB*, 7, 7–9.

Aristotle. *Complete Works.* 2 vols. Edited by Jonathan Barnes. Princeton, N.J., 1984.

Armitage, Angus. *Copernicus and Modern Astronomy.* New York, 2004.

_____. *Sun Stand Thou Still; The Life and Works of Copernicus the Astronomer.* New York, 1947.

Armstrong, A. H. (ed.). *The Cambridge History of Later Greek and Early Medieval Philosophy.* Cambridge, 1967.

Bacon, Francis. *The New Organum and Related Writings.* Edited by Fulton H. Anderson. New York, 1960.

Baker, Robert H. *Astronomy: An Introduction.* New York, 1930.

Bald, R. C. *John Donne, a Life*. Oxford, 1971.

Bede. *The Ecclesiastical History of the English People*. London, 1930.

_____. *The Reckoning of Time*. Liverpool, 2004.

Borndörfer, Rolf, Martin Grötschel, and Andreas Löhal. "Alcuin's Transportation Problems and Integer Programming." Konrad-Zuserum für Infomationsteknik, Berlin, November 1995, pp. 1–3.

Bos, H. J. M. "Huygens, Christiaan," *DSB*, 6, 597–612.

Boyer, Carl B. *A History of Mathematics*. New York, 1968.

Brown, Peter. *The World of Late Antiquity, AD 150–750*. New York, 1971.

Brown, Theodore M. "Descartes: Physiology," *DSB*, 4, 51–65.

Burnet, John. *Greek Philosophy, Thales to Plato*. London, 1950.

Burnett, Charles. *The Introduction of Arab Learning into England*. London, 1997.

Bussard, H. L. L. "Viéte, Françoise," *DSB*, 14, 18–25.

Butterfield, Herbert. *The Origins of Modern Science, 1300–1800*. New York, 1957.

Bylebyl, Jerome J. "Harvey, William," *DSB*, 6, 150–62.

Cahill, Thomas. *How the Irish Saved Civilization*. New York, 1995.

Callus, Daniel A. (ed.). *Robert Grosseteste, Scholar and Bishop*. Oxford, 1955.

Cantor, Norman F. *The Civilization of the Middle Ages*. New York, 1993.

Caspar, Max. *Kepler*. Translated by C. Doris Hellman. New York, 1962.

Chatillon, Jean. "Giles of Rome," *DSB*, 5, 402–3.

Chaucer, Geoffrey. *The Canterbury Tales*. London, 1961.

Clagett, Marshall. "Adelard of Bath," *DSB*, 1, 61–64.

_____, (ed.). *Archimedes in the Middle Ages*, Vol. I: *The Arabo-Latin Tradition*. Madison, Wis., 1964.

_____. *Greek Science in Antiquity*. London, 2001.

_____. "John of Palermo," *DSB*, 7, 133–34.

_____. "Oresme, Nicole," *DSB*, 10, 223–30.

_____. *The Science of Mechanics in the Middle Ages*. Madison, Wis., 1959.

Cobban, Alan B. *The Medieval Universities, Their Development and Organization*. London, 1975.

Cohen, I. Bernard. *Revolution in Science*. Cambridge, Mass., 1985.

Colish, Marcia L. *Medieval Foundations of the Western Intellectual Tradition 400–1400*. New Haven, Conn., and London, 1997.

Collins, Roger. *Early Medieval Europe, 300–1100*. Basingbroke, 1991.

Copernicus, Nicolaus. *De revolutionibus* (*On the Revolutions of the Celestial Spheres*). Translated by Glen Wallis. Edited by Stephen Harking. Philadelphia, 2002.

Cottingham, John. *Descartes*. Oxford, 1986.

_____, (ed.). *The Cambridge Companion to Descartes*. Cambridge, 1992.

Cottingham, John, Robert Stoothoff, and Rugald Murdoch. *The Philosophical Writings of Descartes*. 2 vols. Cambridge, 1985.

Crombie, A. C. "Descartes," *DSB*, 4, 51–55.

_____. "Grosseteste, Robert," *DSB*, 5, 548–54.

_____. *Medieval and Early Modern Science*. 2 vols. 2nd ed. Cambridge, Mass., 1963.

_____. *Robert Grosseteste and the Origins of Experimental Science, 1100–1700*. Oxford, 1953.

_____ (ed.). *Scientific Change; Historical Sketches in the Intellectual, Social and Technical Conditions for Scientific Discovery and Technical Invention from Antiquity to the Present*. New York, 1963.

Crombie, A. C., and J. D. North. "Bacon, Roger," *DSB*, 1, 377–85.

Crosby, H. Lamar (ed.). *Thomas of Bradwardine, His Tractatus de Proportionibus*. Madison, Wis., 1955.

Dales, Richard. *The Scientific Achievements of the Middle Ages*. Philadelphia, 1973.

Daly, John F. "Sacrobosco, Johannes de (John of Holywood)," *DSB*, 12, 60–63.

Dante. *The Divine Comedy*. Translated by H. F. Cary. New York, 1908.

Davies, Norman. *Europe, a History*. Oxford and New York, 1996.

_____. *The Isles, a History*. New York, 1999.

De Santillana, Giorgio. *The Crime of Galileo*. Chicago, 1955.

Descartes, René. *The Philosophical Writings of Descartes*. 3 vols. Translated by John Cottingham, Robert Stoothoff, and Dugald Murdoch. Cambridge, 1985 (vols. 1 & 2), 1991 (vol. 3).

Dictionary of Scientific Biography (DSB). 16 vols. Edited by Charles Coulston Gillispie. New York, 1970–1980.

Donne, John. *The Complete Poetry of John Donne*. Edited by John T. Shawcross. London, 1968.

_____. *The Complete Poetry and Selected Prose of John Donne*. Edited by John Hayward. London, 1929.

Drake, Stillman (trans.). *Discoveries and Opinions of Galileo*. Garden City, N.Y., 1952.

_____. "Benedetti, Giovanni Battista," *DSB*, 1, 604–9.

_____. "Galilei, Galileo," *DSB*, 5, 237–48.

Dreyer, J. L. E. *A History of Astronomy from Thales to Kepler*. New York, 1953.

_____. *Tycho Brahe*. New York, 1963.

Dronke, Peter (ed.). *A History of Twelfth-Century Western Philosophy*. Cambridge, 1988.

Easton, Joy B. "Dee, John," *DSB*, 4, 5–6.

————. "Digges, Thomas," *DSB*, 4, 97–98.

————. "Recorde, Robert," *DSB*, 11, 338–40.

Eicholz, David E. "Pliny," *DSB*, 11, 38–40.

Encyclopedia of the Scientific Revolution. Edited by William Applebaum. New York, 2000.

Ferguson, Kitty. *Tycho and Kepler: The Unlikely Partnership That Forever Changed Our Understanding of the Heavens*. New York, 2002.

Folkerts, Menso. "Regiomontanus' Role in the Transmission and Transformation of Greek Mathematics." In *Tradition, Transmission, Transformation*, pp. 89–113.

Fox, Robert. *Thomas Harriot: An Elizabethan Man of Science*. Farnham, Surrey, 2000.

Freely, John. *Aladdin's Lamp: How Greek Science Came to Europe Through the Islamic World*. New York, 2009.

————. *The Emergence of Modern Science, East and West*. Istanbul, 2004.

Freemantle, Anne. *The Age of Belief: The Medieval Philosophers*. New York, 1954.

Furley, David J. "Lucretius," *DSB*, 8, 536.

Gade, John A. *The Life and Times of Tycho Brahe*. Princeton, N.J., 1947.

Galilei, Galileo. *Dialogue Concerning the Two World Systems, Ptolemaic and Copernican*. Translated by Stillman Drake. Berkeley, Calif., 1967.

————. *Discourses Concerning Two New Sciences, of Mechanics and of Motion*. New York, 1914.

Gassendi, Pierre. *The Life of Copernicus (1473–1543)*. With notes by Olivier Thill. Fairfax, Va., 2003.

Gaukroger, Stephen. *Descartes: An Intellectual Biography*. Oxford, 1995.

_____. *The Emergence of a Scientific Culture: Science and the Shaping of Modernity 1210–1685*. Oxford, 2006.

Geanakoplos, Dino John. *Greek Scholars in Venice: Studies in the Dissemination of Greek Learning from Byzantium to Western Europe*. Cambridge, Mass., 1962.

Geymonat, Ludovico. *Galileo Galilei: A Biography and Inquiry into His Philosophy of Science*. New York, 1965.

Gingerich, Owen. *The Book Nobody Read: Chasing the Revolutions of Nicolaus Copernicus*. New York, 2004.

_____. "Kepler, Johannes," *DSB*, 7, 289–312.

_____. "Reinhold, Erasmus," *DSB*, 11: 365–67.

Glick, Thomas F. "Leo the African," *DSB*, 8, 190–92.

Gliozzi, Marion. "Torricelli, Evangelista," *DSB*, 13, 433–40.

Grant, Edward. *God and Reason in the Middle Ages*. Cambridge, 2001.

_____. "Jordanus de Nemore," *DSB*, 7, 171–79.

_____. "Peregrinus, Peter," *DSB*, 10, 532–40.

_____. *Physical Science in the Middle Ages*. New York, 1971.

Guerlac, Henry. "Copernicus and Aristotle's Cosmos," *Journal of the History of Ideas* 29, no. 1 (1968): 109–13.

Gutas, Dimitri. *Greek Thought, Arabic Culture: The Graeco-Arabic Translation Movement in Baghdad and Early Abbasid Society*. London, 1998.

Guthrie, William K. C. *A History of Greek Philosophy*. 6 vols. Cambridge, 1962–1981.

Hall, A. R. *The Scientific Revolution, 1500–1800*. Boston, 1956.

Hall, Marie Boas. "Boyle, Robert," *DSB*, 2, 377–82.

_____. *Robert Boyle and Seventeenth-Century Chemistry*. Cambridge, 1958.

Hammond, Nicholas (ed.). *The Cambridge Companion to Pascal*. Cambridge, 2003.

Hannam, James. *God's Philosophers: How the Medieval World Laid the Foundations of Modern Science*. London, 2009.

Haring, Nikolaus M. "Thierry of Chartres," *DSB*, 13, 339–41.

Harkness, Deborah E. *The Jewel Box: Elizabethan London and the Scientific Revolution*. New Haven, Conn., 2007.

Haskins, Charles Homer. *The Renaissance of the Twelfth Century*. New York, 1957.

_____. *The Rise of Universities*. New York, 1923.

_____. *Studies in the History of Mediaeval Science*. Cambridge, 1924.

Heath, T. L. *Aristarchus of Samos, the Ancient Copernicus*. Oxford, 1959.

_____. *Diophantus of Alexandria: A Study in the History of Greek Algebra*. New York, 1964.

Hellman, C. Doris. "Brahe, Tycho," *DSB*, 2, 401–14.

Hellman, C. Doris, and Noel M. Swerdlow. "Peurbach, Georg," *DSB*, 15, 473–79.

Heniger, Johann. "Leeuwenhoek, Antoni van," *DSB*, 8, 126–30.

Heninger, S. K. *Touches of Sweet Harmony: Pythagorean Cosmology and Renaissance Poetics*. San Marino, Calif., 1974.

Henry, John. *The Scientific Revolution and the Origins of Modern Science*. New York, 2002.

Hesse, Mary. "Bacon, Francis," *DSB*, 1, 372–77.

Hirst, Anthony, and Michael Silk. *Alexandria, Real and Imagined*. Aldershot, Hampshire, 2004.

Hofmann, Joseph E. "Cusa, Nicholas," *DSB*, 3, 512–16.

_____. "Leibniz, Gottfried Wilhelm," *DSB*, 8, 149–68.

Huff, Toby E. *The Rise of Early Modern Science: Islam, China and the West*. Cambridge, 1993.

Irby-Massie, Georgia L., and Paul T. Keyser. *Greek Science of the Hellenistic Era: A Sourcebook*. London, 2002.

Jayawardene, S. "Pacioli, Luca," *DSB*, 10, 269–72.

Jeanneu, Edouard. "Bernard Silvestre," *DSB*, 2, 21–22.

Jerome. *Selected Letters*. Translated by F. A. Wright. Cambridge, Mass., 1954.

Johnson, Francis R. *Astronomical Thought in Renaissance England*. Baltimore, 1937.

_____. "The Influence of Thomas Digges in the Progress of Modern Astronomy in Sixteenth-Century England," *Osiris* 1 (June 1936): 390–410.

_____. "Marlowe's Astronomy and Renaissance Skepticism," *Journal of English Literary History* 3, no. 4 (December 1946): 241–54.

Jones, Charles W. "Bede, The Venerable," *DSB*, 1, 564–66.

Kantorowicz, Ernst. *Frederick the Second, 1194–1250*. Translated by E. O. Lorimer. New York, 1931.

Keats, John. *The Complete Poems*. Edited by John Barnard. New York, 1988.

Keele, Kenneth D., Ladidlao Reti, Augusto Marinoni, and Marshall Claggett. "Leonardo da Vinci," *DSB*, 8, 192–245.

Kelly, Suzanne. "Gilbert, William," *DSB*, 5, 396–401.

Kirk, G. S., and J. E. Raven. *The Presocratic Philosophers*. Cambridge, 1962.

Kline, Morris. *Mathematical Thought from Ancient to Modern Times*. 3 vols. New York, 1990.

Koestler, Arthur. *The Sleepwalkers: A History of Man's Changing Vision of the Universe*. London, 1959.

Kopal, Zdenek. "Roemer, Olaus," *DSB*, 11, 525–27.

Koran. *The Meaning of the Glorious Koran*. Translated by Mohammed Marmaduke Pickthall. New York, 1953.

Koyré, Alexandre. *The Astronomical Revolution: Copernicus, Kepler, Borelli*. Translated by R. E. W. Madison. Ithaca, N.Y., 1973.

————. *From the Closed World to the Infinite Universe*. New York, 1958.

————. *Newtonian Studies*. Cambridge, Mass., 1965.

Krafft, Fritz. "Guericke, Otto von," *DSB*, 5, 574–76.

Kren, Claudia. "Alain de Lille," *DSB*, 1, 91–92.

————. "Gundissalinus, Dominicus," *DSB*, 5, 591–93.

————. "Hermann the Lame," *DSB*, 6, 301–3.

————. "Roger of Hereford," *DSB*, 11, 503–4.

Kuhn, Thomas S. *The Copernican Revolution: Planetary Astronomy in the Development of Western Thought*. Cambridge, Mass., 1957.

————. *The Structure of Scientific Revolutions*. Chicago, 1976.

Leff, Gordon. "Duns Scotus, John," *DSB*, 4, 254–56.

Lemay, Richard. "Gerard of Cremona," *DSB*, 15, 173–92.

Levey, Martin. "Abraham Bar Hiyya Ha-Nasi (Savasorda)," *DSB*, 1, 22–23.

————. "Ezra, Abraham Ibn," *DSB*, 4, 502–3.

Levy, Tony. "Hebrew Mathematics in the Middle Ages." In *Tradition, Transmission, Transformation*, pp. 71–88.

Lindberg, David C. *The Beginnings of European Science: The European Scientific Tradition in Philosophical, Religious and Institutional Context, 600 BC to AD 1450*. Chicago, 1992.

————. "Pecham, John," *DSB*, 10, 473–76.

———— (ed.). *Science in the Middle Ages*. Chicago, 1978.

————. *Studies in the History of Medieval Optics*. London, 1983.

_____. *Theories of Vision from al-Kindi to Kepler.* Chicago, 1976.

_____. "Witelo," *DSB*, 14, 457–62.

Lindberg, David C., and Robert S. Westman (eds.). *Reappraisal of the Scientific Revolution.* Cambridge, 1990.

Little, A. G. (ed.). *Roger Bacon Essays.* Oxford, 1914.

Lloyd, G. E. R. *Early Greek Science, Thales to Aristotle.* London, 1970.

_____. *Greek Science After Aristotle.* London, 1973.

Lohne, J. A. "Harriot, Thomas," *DSB*, 6, 124–29.

Lorris, Guilaume de, and Jean de Meun. *The Romance of the Rose.* Translated by Harry W. Robbins. New York, 1962.

Lucretius. *De rerum natura (On the Nature of the Universe).* Translated by W. H. D. Rouse. London and Cambridge, Mass., 1937.

Lundquist, Stig (ed.). *Nobel Lectures in Physics, 1971–1980.* Göteberg, 1992.

Lyons, Jonathan. *The House of Wisdom: How the Arabs Transformed Western Civilization.* London, 2009.

Machamer, Peter (ed.). *The Cambridge Companion to Galileo.* Cambridge, 1996.

MacLeod, Roy (ed.). *The Library of Alexandria, Centre of Learning in the Ancient World.* London, 2000.

Mahoney, Michal S. "Descartes: Mathematics and Physics," *DSB*, 4, 55–61.

_____. "Fermat, Pierre de," *DSB*, 4, 566–76.

Makdisi, George. *The Rise of Colleges: Institutions of Learning in Islam and the West.* Edinburgh, 1981.

_____. *The Rise of Humanism in Classical Islam and the West.* Edinburgh, 1990.

Manuel, Frank E. *A Portrait of Isaac Newton.* Cambridge, Mass., 1968.

Marlowe, Christopher. *The Complete Plays*. Edited by Frank Romany and Robert Lindsey. London, 2003.

Masotti, Arnaldi. "Maurolico, Francesco," *DSB*, 9, 190–94.

Masson, Georgina. *Frederick II of Hohenstaufen, a Life*. London, 1957.

Matthew, Donald. *The Norman Kingdom of Italy*. Cambridge, 1992.

McCluskey, Stephen C. *Astronomies and Cultures in Early Medieval Europe*. Cambridge, 1998.

McVaugh, Michael. "Constantine the African," *DSB*, 3, 393–95.

Menocal, Maria Rosa. *The Ornament of the World: How Muslims, Jews, and Christians Created a Culture of Tolerance in Medieval Spain*. Boston, 2002.

Milton, John. *Paradise Lost*. Edited by Edward Le Comte. New York, 1961.

_____. *Paradise Regained*. Edited by Merritt Y. Hughes. New York, 1937.

Minio-Paluello, Lorenzo. "Boethius," *DSB*, 2, 228–36.

_____. "James of Venice," *DSB*, 7, pp. 65–67.

_____. "Michael Scott," *DSB*, 9, 361–65.

_____. "Moerbeke, William of," *DSB*, 9, 434–40.

_____. "Plato of Tivoli," *DSB*, 11, 31–23.

Mitchell, Stephen. *A History of the Later Roman Empire AD 284–641: The Transformation of the Ancient World*. Oxford, 2007.

Molland, A. C. "John of Dumbleton," *DSB*, 7, 116–17.

Monfasani, John. *Byzantine Scholars in Renaissance Italy: Cardinal Bessarion and Other Emigres; Selected Essays*. Brookfield, Vt., 1995.

_____. *Greeks and Latins in Renaissance Italy: Studies on Humanism and Philosophy in the 15th Century*. Brookfield, Vt., 2004.

Montgomery, Scott L. *Science in Translation: Movements of Knowledge Through Cultures and Time*. Chicago, 2000.

Moody, Ernest A. "Albert of Saxony," *DSB*, 1, 93–95.

————. "Buridan, Jean," *DSB*, 2, 603–8.

————. "Galileo and Avempace: The Mechanics of the Leaning Tower Experiment," *Journal of the History of Ideas* 12 (1951): 375–422.

————. "Ockham, William of," *DSB*, 10, 223–30.

Murdoch, John. "Bradwardine, Thomas," *DSB*, 2, 390–97.

Murdoch, John, and Edith Dudley Sylla. "Burley, Walter," *DSB*, 2, 608–12.

————. "Swineshead, Richard," *DSB*, 13, 184–213.

Netz, Reviel, and William Noel. *The Archimedes Codex: Revealing the Blueprint of Modern Science*. London, 2007.

Newton, Isaac. *Opticks, or a Treatise on the Reflections, Refractions, Inflections, and Colours of Light*. London, 1952.

————. *Principia (Mathematical Principles of Natural Philosophy)*. Translated by I. Bernard Cohen and Anne Whitman. Berkeley, Calif., 1999.

North, John David. *The Norton History of Astronomy and Cosmology*. New York, 1995.

————. "Richard of Wallingford," *DSB*, 11, 414–16.

O'Leary, De Lacy. *How Greek Science Passed to the Arabs*. London, 1949.

O'Neil, W. M. *Early Astronomy from Babylonia to Copernicus*. Sydney, 1986.

————. *Time and the Calendar*. Sydney, 1975.

Osler, Margaret (ed.). *Rethinking the Scientific Revolution*. New York, 2000.

Paluello, L. Minio. "Pierre Abailard (Peter Abelard)," *DSB*, 11, 1–4.

Pannekoek, Anton. *A History of Astronomy*. New York, 1961.

Payne-Gaposchkin, Cecelia. *Introduction to Astronomy*. New York, 1954.

Permuda, Loris. "Abano, Pietro d'," *DSB*, 1, 4–5.

Pingree, David. "Leo the Mathematician," *DSB*, 8, 190–92.

Plato. *Complete Works*. Edited by John M. Cooper. Indianapolis, 1997.

Pliny the Elder. *Natural History*. 10 vols. Translated by H. Rackham et al. Cambridge, Mass., 1942–1963.

Poulle, Emmanuel. "John of Saxony," *DSB*, 7, 139–41.

————. "John of Sicily," *DSB*, 7, 141–42.

————. "William of St. Cloud," *DSB*, 14, 389–91.

Ragep, F. Jamil. "Ali Qushji and Regiomontanus: Eccentric Transformations and Copernican Revolutions," *Journal for the History of Astronomy* 36 (2005): 359–71.

————. "Copernicus and His Islamic Predecessors: Some Historical Remarks," *Filozofski vestnik* 25, no. 2 (2004): 125–42.

Ragep, F. Jamil, and Sally P. Ragep with Steven Livesey (eds.). *Tradition, Transmission, Transformation*. Leiden, 1996.

Rashdall, Hastings. *The Universities in Europe in the Middle Ages*. 3 vols. London, 1936.

Rochot, Bernard. "Gassendi, Pierre," *DSB*, 5, 284–90.

Ronan, Colin A. *The Cambridge Illustrated History of the World's Science*. London, 1983.

————. "Halley, Edmond," *DSB*, 6, 67–74.

Rosen, Edward. "Commandino, Federico," *DSB*, 3, 363–65.

————. "Copernicus, Nicholas," *DSB*, 3, 401–11.

————. "Mastlin, Michael," *DSB*, 9, 167–71.

————. "Osiander, Andreas," *DSB*, 10, 245–46.

_____. "Regiomontanus, Johannes," *DSB*, 11, 348–52.

_____. "Rheticus, George Joachim," *DSB*, 11, 395–98.

_____. *Three Copernican Treatises*. New York, 1959.

_____. "Was Copernicus a Neoplatonist?," *Journal of the History of Ideas* 44, no. 4 (October 1983): 667–69.

Rosinska, Grazyna. "Nasir al-Din al-Tusi and Ibn al-Shatir in Cracow?," *Isis* 65, no. 2 (1974): 239–43.

Rubenstein, Richard E. *Aristotle's Children: How Christians, Muslims, and Jews Rediscovered Ancient Wisdom and Illuminated the Middle Ages*. New York, 2001.

Rudnicki, Jozef. *Nicolaus Copernicus*. London, 1943.

Runciman, Steven. *Byzantium and the Renaissance*. Tucson, 1970.

_____. *The Last Byzantine Renaissance*. Cambridge, 1970.

Sabra, A. I. "The Appropriation and Subsequent Naturalization of Greek Science in Medieval Islam: A Preliminary Statement." In *Tradition, Transmission, Transformation*, pp. 3–27.

_____. *Theories of Light, from Descartes to Newton*. London, 1967.

Sa'di of Shiraz. *Tales from the Bustun*.... Translated by Reuben Levy. London, 1928.

Saliba, George. *Islamic Science and the Making of the European Renaissance*. Cambridge, Mass., 2007.

Sambursky, Samuel. *The Physical World of the Greeks*. London, 1956.

_____. *Physics of the Stoics*. New York, 1959.

_____. "John Philoponus," DSB 7, 135-6

Samso, Julio. "Levi ben Gerson," *DSB* 8, 279–82.

Santillana, Giorgio de. *The Crime of Galileo*. Chicago, 1955.

Sarton, George. *The Appreciation of Ancient and Medieval Science During the Renaissance, 1459–1600*. Philadelphia, 1955.

_____. *A History of Science*. 2 vols. Cambridge, Mass., 1952, 1959.

_____. *Introduction to the History of Science*. 3 vols. in 5 parts. Baltimore, 1927–1948.

Schneidler, F. "Schöner, Johannes," *DSB*, 12, 199–200.

Scott, J. F. "Wren, Christopher," *DSB*, 14, 509–11.

Seneca. *Natural Questions*. Translated by J. Clarke. London, 1910.

Settle, Thomas W. "Borelli, Giovanni Alfonso," *DSB*, 2, 306–14.

Shakespeare, William. *The Complete Works*. Edited by Peter Alexander. London, 1951.

Shank, Michael H. "The Classical Scientific Tradition in Fifteenth-Century Vienna." In *Tradition, Transmission, Transformation*, pp. 115–36.

Shapin, Steven. *The Scientific Revolution*. Chicago, 1996.

Sharpe, William D. "Isidore of Seville," *DSB*, 7, 28–30.

Singer, Charles, E. J. Holmyard, and A. R. Hall. *A History of Technology*. 2 vols. Oxford, 1954–1984.

Siraisi, Nancy. *Medieval and Early Renaissance Medicine: An Introduction to Knowledge and Practice*. Chicago, 1990.

Spenser, Edmund. *The Faerie Queene*. 2 vols. New York, 1927.

Stahl, William H. "The Greek Heliocentric Theory and Its Abandonment," *Transactions and Proceedings of the American Philological Society* 76 (1945): 321–32.

_____. "Macrobius," *DSB*, 9, 1–2.

_____. "Martianus Capella," *DSB*, 9, 140–41.

Steel, Duncan. *Marking Time: The Epic Quest to Invent the Perfect Calendar*. New York, 2000.

Stern, S. M. "Isaac Israeli," *DSB*, 7, 22–23.

Strabo. *The Geography*. 8 vols. Translated by Horace Leonard Jones. Cambridge, Mass., 1969.

Struik, D. J. "Gerbert d'Aurillac," *DSB*, 5, 364–66.

Sullivan, J. W. N. *Isaac Newton, 1642–1727*. New York, 1928.

Swerdlow, N. M., and O. Neugebauer. *Mathematical Astronomy in Copernicus' De Revolutionibus*. 2 vols. New York, 1984.

Talbot, Charles H. "Stephen of Antioch," *DSB*, 13, 38–39.

Taton, René. *History of Science*. 4 vols. Translated by A. J. Pomerans. New York, 1964–1966.

_____. "Pascal, Blaise," *DSB*, 10, 330–42.

Thomas, Phillip Drennon. "Alcuin of York," *DSB*, 1, 104–5.

_____. "Alfonso el Sabio," *DSB*, 1, 122.

_____. "Cassiodorus," *DSB*, 3, 109–10.

Thoren, Victor E. "Flamsteed, John," *DSB*, 5, 22–26.

Thorndike, Lynn. *A History of Magic and Experimental Science*. 8 vols. New York, 1923–1956.

_____. *Michael Scot*. London, 1966.

Thucydides. *History of the Peloponnesian War*. Translated by Rex Warner. Harmondsworth, 1987.

Toomer, G. J. "Campanus of Novara," *DSB*, 3, 23–29.

Urquhart, John. "How Islam Changed Medicine ...," *British Medical Journal* 332 (January 14, 2006): 120.

Van Helden, Albert. *Measuring the Universe: Cosmic Dimensions from Aristarchus to Halley*. Chicago, 1985.

Vescovini, Graziella, Federic. "Francis of Meyronnes," *DSB*, 5, 115–17.

Vitruvius. *The Ten Books on Architecture (De Architectura)*. Translated by Morris Hicky Morgan. New York, 1966.

Vogel, Kurt. "Byzantine Science." In *Cambridge Medieval History*, new edition, vol. 4, part 2, pp. 264–305.

————. "Fibonacci, Leonardo (Leonardo of Pisa)," *DSB*, 4, 604–13.

Wallace, William A. "Albertus Magnus, Saint," *DSB*, 1, 99–103.

————. "Aquinas, Saint Thomas," *DSB*, 1, 196–200.

————. "Dietrich of Freiberg," *DSB*, 4, 92–5.

————. "Galileo's Pisan Studies in Science and Philosophy." In Peter Machamer (ed.), *The Cambridge Companion to Galileo*, pp. 27–52.

————. "William of Auvergne," *DSB*, 14, 388–89.

Ward, Benedicta. *The Venerable Bede*. Harrisburg, Pa., 1990.

Ward-Perkins, Brian. *The Fall of Rome and the End of Civilization*. Oxford and New York, 2006.

Watt, W. Montgomery. *The Influence of Islam on Medieval Europe*. Edinburgh, 1982.

Webb, J. F. (trans.). *The Age of Bede*. New York, 1988.

Westfall, Richard S. *Never at Rest: A Biography of Isaac Newton*. Cambridge, 1980.

Whitfield, Peter. *Landmarks in Western Science: From Prehistory to the Atomic Age*. London, 1991.

Wilson, Curtis A. "Heytesbury, William," *DSB*, 6, 376–80.

Wilson, Nigel. *From Byzantium to Italy: Greek Studies in the Italian Renaissance*. London, 1992.

Wipple, John F., and Allan B. Walker. *Medieval Philosophy from St. Augustine to Nicholas of Cusa*. New York, 1960.

Yates, Frances A. "Bruno, Giordano," *DSB*, 2, 539–44.

————. *Giordano Bruno and the Hermetic Tradition*. Chicago, 1991.

Youschevitch, A. P. "Newton, Isaac," *DSB*, 10, 42–103.

Acknowledgments

I would like to thank the staff of the Bosphorus University Library for all of the assistance they have given me in my research for this book. I would also like to thank my editor, Dan Crissman, for his invaluable help in preparing this manuscript for publication.

Index

Gibbon, Edward, 44
Giese, Tiedemann, 229, 230
Gilbert, William, 143, 244
Gingerich, Owen, 236, 237, 258
Giudecci, Mario, 282
grammar, 43, 54, 60, 104, 161, 206
Grant, Edward, 145
Grassi, Horatio, 282
gravity, 144, 145, 162, 165, 166,
    169, 171, 172, 224, 231, 264,
    268, 270, 271, 284, 298, 299,
    303, 307, 309, 310
Greatorex, Ralph, 300
Grosseteste, Robert, 109, 125, 227;
    and Aristotle, commentaries
    on, 123; and Bacon, 131–37;
    calendar reform, 129–30; in-
    fluence of, 137, 140, 147,
    150, 153, 175–78, 227; and
    mathematics, 124; *On the
    Fixity of Motion and Time*,
    128; and optics, 124, 126–28,
    175–78, 184–86; *Rainbow*,
    128; and refraction, 127; sci-
    entific method, 123–24; *De
    sphaera*, 129, 194–95
Guericke, Otto von, 300
Gutenberg Bible, 217
Guthrie, William, 99

Hadrian the African, 52
Hakem II, al-, 80
Haller, Johann, 218

Halley, Edmond, 261, 306–7, 311
Haring, Nikolaus, 100
Harriot, Thomas, 257, 259, 279
Hasib, Habash al-, 69
Haskins, Charles Homer, 84, 98,
    102
Haytham, Abu 'Ali al-Hasan ibn
    al- (Alhazen), 74, 83, 90, 177,
    178, 180–86
Hegel, Georg Wilhelm Friedrich,
    62
Helbing, Mario, 267
Helden, Albert van, 198
Hellman, Doris, 214
Henricus Aristippus, 95
Heraclides Ponticus, 29, 45, 46
Heraclitus of Ephesus, 19–20
Hermann the Lame, 75
Hero of Alexandria, 36, 71, 95, 96,
    177, 178,
Heytesbury, William, 153, 157–59
Hikma, Bayt al-, 67, 70
Hipparchus of Nicaea, 36, 38, 42,
    63, 72, 75, 110, 129, 194,
    267
Hippocrates, 26, 48, 49, 70, 94,
    104
*Historia ecclestiastica gentis Anglo-
    rus (The Ecclesiastical His-
    tory of the English People)*, 53
homocentric spheres, 26, 28, 45,
    117, 223, 245
Hooke, Robert, 300, 305–6

Sa'di of Shiraz, 72
Sacrobosco, Johannes de, 147,
    195–97, 199, 218, 245
Saffah, Abu al-Abbas al-, 80
Sagredo, Giovan Francesco, 284
Salviati, 283, 285, 288
Sandivogius of Czecel, 218
Sarpi, Paolo, 270
Sarsi, Lotario, 282, 283
Savasorda. *See* Ha-Nasi
Scheiner, Christopher, 278–79
Schöner, Johann, 228
Schopenhauer, Arthur, 62
Scientific Revolution, 11, 227, 266,
    277, 291, 314
Scotus, John Duns, 147
Second Council of Lyons, 95, 107
sector, 270
Seneca the Younger, 44, 218
seven liberal arts, 43, 46, 58–62,
    97, 122
sextant, 246
Shapin, Steven, 291
Siger of Brabant, 115
Silvestre, Bernard, 101, 314
Snellius, Willebrord, 266
Socrates, 13, 22, 23
Soto, Domingo de, 271
Spenser, Edmund, 21
Stahl, William H., 45
Stephen of Antioch, 76–77, 103
Stevin, Simon, 265–66
Stoics, 30–31, 100

Strabo, 50
Strato of Lampsacus, 31
Swerdlow, Noel, 214
Swineshead, Richard, 153, 160
Synod of Whitby, 53, 57

tables, astronomical (*ephemerides*),
    80, 84, 193, 197, 211, 239;
    of Adelard, 75, 84; *Alfonsine
    Tables*, 202, 214, 218, 227,
    245; of Bede, 54, 58; and
    Brahe, 245, 254; and Cam-
    panus, 198; and Copernicus,
    218, 227, 233, 239;
    Ephemeris (Feild), 240–41; of
    Grosseteste, 129, 135; of al-
    Hasib, 69; of al-Khwarizmi,
    69, 77, 78, 84; *Persian
    Tables*, 207; of Peurbach,
    210, 214, 218; *Prutenic Ta-
    bles* (Reinhold), 239, 240,
    245; of Ptolemy, 36; and Re-
    giomontanus, 210, 211, 213,
    214; *Rudolphine Tables* (Ke-
    pler), 254, 256, 258, 260;
    Tabulae directionum, 211,
    218; *Tabulae eclipsium,* 210,
    214, 218; *Toledan Tables
    (Marseille Tables)*, 198–201;
    of Walcher, 75
telescope, 110, 121, 134, 244, 257,
    258, 271, 272, 274, 275, 276,
    278, 304, 312